Genome Mining and Synthetic Biology in Marine Natural Products Discovery

Genome Mining and Synthetic Biology in Marine Natural Products Discovery

Editor

Maria Costantini

MDPI • Basel • Beijing • Wuhan • Barcelona • Belgrade • Manchester • Tokyo • Cluj • Tianjin

Editor
Maria Costantini
Stazione Zoologica Anton Dohrn
Italy

Editorial Office
MDPI
St. Alban-Anlage 66
4052 Basel, Switzerland

This is a reprint of articles from the Special Issue published online in the open access journal *Marine Drugs* (ISSN 1660-3397) (available at: https://www.mdpi.com/journal/marinedrugs/special_issues/gnm_mn_syn_bio).

For citation purposes, cite each article independently as indicated on the article page online and as indicated below:

LastName, A.A.; LastName, B.B.; LastName, C.C. Article Title. *Journal Name* **Year**, *Volume Number*, Page Range.

ISBN 978-3-0365-0254-0 (Hbk)
ISBN 978-3-0365-0255-7 (PDF)

Contents

About the Editor

Maria Costantini is a molecular biologist. Her research is focused on studying secondary metabolites from marine organisms as a source of stress as well as biologically and pharmacologically active compounds, using molecular biology approaches.

The aims of her current research in molecular ecology include:

- Studying the molecular basis of cellular mechanisms that underlie the responses of benthic organisms in environmental stress: the molecular effects of diatom-derived oxylipins and benthic diatoms on embryonic and larval development using the sea urchin as a model system;

- Studying the potential effects of crude extract from marine organisms (such as sponges) with the aim to isolate biologically active compounds with pharmacological application and new molecules for drug discovery.

 *marine drugs*

Editorial

Genome Mining and Synthetic Biology in Marine Natural Product Discovery

Maria Costantini

Department of Marine Biotechnology, Stazione Zoologica Anton Dohrn, Villa Comunale, 80121 Napoli, Italy; maria.costantini@szn.it; Tel.: +39-081-583-3285

Received: 17 November 2020; Accepted: 22 November 2020; Published: 3 December 2020

Oceans cover nearly 70% of the earth's surface and host a huge ecological, chemical, and biological diversity. The natural conditions of the sea favor, in marine organisms, the production of a large variety of novel molecules with great pharmaceutical potential. Marine organisms are unique in their structural and functional features compared to terrestrial ones [1,2]. Innovation in this field is very rapid, as revealed by the funding of several Seventh Framework Programme (FP7) and Horizon 2020 projects under the topic "Blue Growth". The major parts of these projects have the common final goal of meeting the urgent need to discover new drug entities to counteract the increased incidence of severe diseases (such as cancer) and the reduced efficacy of existing drugs [3].

For this reason, the application of molecular biology techniques in biotechnological field is very important. Driven by the rapidly-decreasing cost and increasing throughput of DNA-sequencing technologies, significant progress in genomics has renewed interest in the discovery of natural products. Rapidly-expanding genomic and metagenomic datasets reveal a vast number of biosynthetic gene clusters (BGCs) in nature, which are predicted to encode novel biomedically-relevant molecules. This 'top-down' discovery approach can only provide access to a small fraction of biosynthetic compounds, given that the majority of microorganisms cannot be isolated or cultured. Furthermore, even in culturable organisms, the encoded secondary metabolites of many biosynthetic gene clusters are unknown. This may be due to strong down-regulation of product biosynthesis at the transcriptional, translational, and/or post-translational levels in the absence of the right activating cues in the laboratory. Alternatively, secondary metabolites that are produced at very low yields may escape detection and characterization. Natural product discovery, which used to be predominantly an analytical chemistry problem, has become a challenge in genomics and molecular biology with respect to how to manipulate relevant genes and sequences to produce the desired encoded metabolites.

In recent years, marine genomics has grown rapidly, helped by the large amount of information that is are becoming available to the international scientific community. Taking into account the current excitement in the field of marine biotechnology, we have assembled this Special Issue entitled "Genome Mining and Synthetic Biology in Marine Natural Product Discovery". The aim of this Special Issue is to assess the impact of these molecular approaches on the discovery of bioactive compounds from marine organisms. The term "genome mining" is used to identify all bioinformatic investigation aimed at detecting the biosynthetic pathways of bioactive natural products and their possible functional and chemical interactions [4]. Genome mining includes the identification of BGCs of previously-uncharacterized natural products within the genomes of sequenced organisms, sequence analysis of the enzymes encoded by these gene clusters, and the experimental identification of the products of the gene clusters [5].

This special issue includes one review article and four original studies on the application of genome mining and synthetic biology in promoting marine natural products for biotechnological applications.

An up-to-date view on the importance of genome mining to identify new natural products is provided in the review by Albarano et al. [6] In the first part of this review, the significance of genome mining and other associated techniques is discussed, as well as their possible advantages and disadvantages. One of the strengths of using genome mining is the detection of a large amount of bioactive natural compounds. Genome mining has the advantages of being relatively cheap and easy to apply in the laboratory and not requiring particular skills and/or experience by the operators. The weaknesses of this bioinformatic approach are that: (1) only known biosynthetic gene clusters can be identified and (2) it is not possible to predict the biotechnological potential of the natural products under analysis, as well as to formulate their chemical structures. In addition, the authors include not only marine organisms but also terrestrial ones, considering that molecular approaches were only recently applied to the discovery of bioactive compounds from the sea.

A very important point arising from this review is that the available literature on genome mining mainly concerns bacteria, from which the majority of new natural drugs are isolated. In fact, Gao et al. [7] report on the cloning, recombinant expression, and characterization of the ulvan lyase (able to degrade ulvan to oligosaccharides) gene, *ALT3695*, from marine bacterium *Alteromonas* sp. A321. Its amino acid sequence exhibits low sequence identity with the polysaccharide lyase family 25 (PL25) ulvan lyases from *Pseudoalteromonas* sp. PLSV (64.14%), *Alteromonas* sp. LOR (62.68%), and *Nonlabens ulvanivorans* PLR (57.37%). Ulvan is a type of cell wall polysaccharide, extracted from marine green seaweed (belonging to the genera *Ulva* and *Enteromorpha*), with a great biotechnological potential thanks to its various pharmacological properties including antioxidant, anticoagulant, antitumor, antihyperlipidemic, and antiviral activities. Furthermore, ulvan-derived oligosaccharides has applications in the functional food and pharmaceutical industries. ALT3695 exhibits excellent stability and substrate affinity, showing that it could be used for the preparation of ulvan oligosaccharides.

Other important sources of natural drugs are the fungi. Examples of such drugs are penicillin, cephalosporin, ergotrate, and the statins. Guided by bioinformatic analysis of the genomic sequence of a marine-derived fungal strain of *Penicillium dipodomyis* YJ-11, Yu et al. [8] report for the first time on the application of global regulator LaeA in *P. dipodomyis* (PdLaeA), aiming to the increase the secondary metabolite. In fact, this regulator is able to influence fungi in many aspects of their life, such as increasing or reducing secondary metabolite production, activating cryptic gene clusters, asexual and sexual differentiation, sporulation, and pigmentation. The overexpression of the PdLaeA gene in *P. dipodomyis* YJ-11 lead to the discovery of two new compounds, 10,11-dihydrobislongiquinolide and 10,11,16,17-tetrahydrobislongiquinolide, together with four known sorbicillin analogues.

Microalgae are eukaryotic photosynthetic microorganisms, which can be cultivated in large quantities, and for this reason they are receiving increasing attention from biotechnologists. They are able to produce a wide diversity of species and natural products with pharmaceutical, nutraceutical, and cosmeceutical applications. Riccio et al. [9] investigated the presence of genes involved in monogalactosyldiacylglycerols (MGDGs) and sulfoglycolipid synthesis in microalgae, considering all the microalgal transcriptomes and metatranscriptomes available from the Marine Microbial Eukaryote Transcriptome Sequencing Project (MMETSP) and the recent Tara Oceans and Global Ocean sampling expedition. In particular, the presence of MGDG synthase (MGD), UDP-sulfoquinovose synthase (SQD1), and sulfoquinovosyltransferase (SQD2) were identified in different microalgal classes. Not all the sequences are expressed within each taxa analyzed, suggesting the absence of a specific gene. This study prompted the discovery of new bioactive species with new insights into enzyme discovery from microalgae for biotechnological applications.

Zhu et al. [10] performed research on the genome of *Thraustochytriidae* sp. SZU445, a single-celled eukaryotic marine protist belonging to the class Labyrinthulomycetes. Using a third-generation sequencing platform, fatty acid synthesis pathways, specifically for docosapentaenoic acid (DHA) docosahexaenoic acid (DPA) were identified, predicting and analyzing the Gene Ontology (GO) and Kyoto Encyclopedia of Genes and Genomes (KEGG) databases. This represents a very important step for the scientific community because these marine protists can be cultivated to yield high biomasses containing substantial quantities of lipids abundant in polyunsaturated fatty acids (PUFAs), and so also accumulating large amounts of DHA.

Even though these new molecular approaches need to be ecxpanded, this Special Issue of Marine Drugs provides an interesting panorama of current research on genome mining and other associated experimental techniques. As Guest Editor, I trust that this collection will help the scientific community to inspire the next generation of marine biotechnologies. In fact, molecular and bioinformatic approaches permit us to overcome the over-utilization of marine resources and the use of destructive collection practices, and to apply a more environmentally-friendly approach to drug discovery

Funding: This research received no external funding.

Conflicts of Interest: The author declares no conflict of interest.

References

1. Jimeno, J.; Faircloth, G.; Fernandez Sousa-Faro, J.M.; Scheuer, P.; Rinehart, K. New marine derived anticancer therapeutics. A journey from the sea to clinical trials. *Mar. Drugs* **2004**, *2*, 14–29. [CrossRef]
2. Kijjoa, A.; Sawangwong, P. Drugs and cosmetics from the sea. *Mar. Drugs* **2004**, *2*, 73–82. [CrossRef]
3. Romano, G.; Costantini, M.; Sansone, C.; Lauritano, C.; Ruocco, N.; Ianora, A. Marine microorganisms as a promising and sustainable source of bioactive molecules. *Mar. Environ. Res.* **2017**, *128*, 58–69. [CrossRef] [PubMed]
4. Ziemert, N.; Alanjary, M.; Weber, T. The evolution of genome mining in microbes—A review. *Nat. Prod. Rep.* **2016**, *33*, 988–1005. [CrossRef] [PubMed]
5. Trivella, D.B.B.; De Felicio, R. The Tripod for Bacterial Natural Product Discovery: Genome Mining, Silent Pathway Induction, and Mass Spectrometry-Based Molecular Networking. *mSystems* **2018**, *3*, e00160-17. [CrossRef] [PubMed]
6. Albarano, L.; Esposito, R.; Ruocco, N.; Costantini, M. Genome mining as new challenge in natural products discovery. *Mar. Drugs* **2020**, *18*, 199. [CrossRef]
7. Gao, J.; Du, C.; Chi, Y.; Zuo, S.; Ye, H.; Wang, P. Cloning, Expression, and Characterization of a New PL25 Family Ulvan Lyase from Marine Bacterium *Alteromonas* sp. A321. *Mar. Drugs* **2019**, *17*, 568. [CrossRef]
8. Yu, J.; Han, H.; Zhang, X.; Ma, C.; Sun, C.; Che, Q.; Gu, Q.; Zhu, T.; Zhang, G.; Li, D. Discovery of Two New Sorbicillinoids by Overexpression of the Global Regulator LaeA in a Marine-Derived Fungus *Penicillium dipodomyis* YJ-11. *Mar. Drugs* **2019**, *17*, 446. [CrossRef] [PubMed]
9. Riccio, G.; De Luca, D.; Lauritano, C. Monogalactosyldiacylglycerol and Sulfolipid Synthesis in Microalgae. *Mar. Drugs* **2020**, *18*, 237. [CrossRef]
10. Zhu, X.; Li, S.; Liu, L.; Li, S.; Luo, Y.; Lv, C.; Wang, B.; Cheng, C.H.K.; Chen, H.; Yang, X. Genome Sequencing and Analysis of *Thraustochytriidae* sp. SZU445 Provides Novel Insights into the Polyunsaturated Fatty Acid Biosynthesis Pathway. *Mar. Drugs* **2020**, *18*, 118. [CrossRef] [PubMed]

Publisher's Note: MDPI stays neutral with regard to jurisdictional claims in published maps and institutional affiliations.

 *marine drugs*

Review

Genome Mining as New Challenge in Natural Products Discovery

Luisa Albarano [1,2,†], Roberta Esposito [1,2,†], Nadia Ruocco [1] and Maria Costantini [1,*]

[1]   Department of Marine Biotechnology, Stazione Zoologica Anton Dohrn, Villa Comunale, 80121 Napoli, Italy; luisa.albarano@szn.it (L.A.); roberta.esposito@szn.it (R.E.); nadia.ruocco@gmail.com (N.R.)

[2]   Department of Biology, University of Naples Federico II, Complesso Universitario di Monte Sant'Angelo, Via Cinthia 21, 80126 Napoli, Italy

*   Correspondence: maria.costantini@szn.it; Tel.: +39-081-583-3315; Fax: +39-081-764-1355

†   These authors contributed equally to this work.

Received: 4 March 2020; Accepted: 3 April 2020; Published: 9 April 2020

Abstract: Drug discovery is based on bioactivity screening of natural sources, traditionally represented by bacteria fungi and plants. Bioactive natural products and their secondary metabolites have represented the main source for new therapeutic agents, used as drug leads for new antibiotics and anticancer agents. After the discovery of the first biosynthetic genes in the last decades, the researchers had in their hands the tool to understand the biosynthetic logic and genetic basis leading to the production of these compounds. Furthermore, in the genomic era, in which the number of available genomes is increasing, genome mining joined to synthetic biology are offering a significant help in drug discovery. In the present review we discuss the importance of genome mining and synthetic biology approaches to identify new natural products, also underlining considering the possible advantages and disadvantages of this technique. Moreover, we debate the associated techniques that can be applied following to genome mining for validation of data. Finally, we review on the literature describing all novel natural drugs isolated from bacteria, fungi, and other living organisms, not only from the marine environment, by a genome-mining approach, focusing on the literature available in the last ten years.

Keywords: bacteria; fungi; genome mining; natural products; synthetic biology

---

1. Introduction on Bioactive Natural Products Isolation

Nature is an important source of bioactive products and their derivatives (secondary metabolites), which form part of many important drugs formation widely used in the clinic field [1]. In fact, as reported in Newman and Cragg [2], over the last 30 years the great majority of anticancer, anti-infective, and anti-bacterial drugs are represented by natural products and their derivatives, produced by all organisms (from bacteria to plants, invertebrate, and other animals) with different chemical structure and leading to several biological activities [3,4]. Furthermore, these secondary metabolites have influenced the development of several drugs, including antibacterial, anticancer, and anti-cholesterol agents [5]. Several of these bioactive products are derived from microorganisms, such as fungi and bacteria [6], which have represented an important source of antibiotics and many other medicines [7,8]. In particular many bacteria deriving from the marine environment, particularly those found in association with marine invertebrates (such as sponges), are able to produce secondary metabolites with potential anticancer and antifungal roles because of their cytotoxic properties [9,10]. Considering the great problem of the antimicrobial resistance increase and its high impact on human health, there is an important need of searching for new natural products that could therefore remedy this issue [11,12]. For these reasons, in the past decade, genomic science has been used to identify the possible drug targets and to find novel genes cluster for the biosynthesis of natural products [13]. The development of the genome sequencing technologies to find novel metabolites has surely drown the attention of pharmaceutical industries, which had by now lost interest in natural products due to the advent of combinatorial chemistry [14]. The advent of based-genome sequencing techniques, especially with establishment of genome mining, has allowed to obtain new natural drugs in a faster and cheaper way.

Genome Mining

The term "genome mining" are associated to every bioinformatics investigation used to detect not only the biosynthetic pathway of bioactive natural products, but also their possible functional and chemical interactions [15]. Specifically, the genome mining involves the identification of previously uncharacterized natural product biosynthetic gene clusters within the genomes of sequenced organisms, sequence analysis of the enzymes encoded by these gene clusters, together with the experimental identification of the products of the gene clusters (Figure 1; [16]).

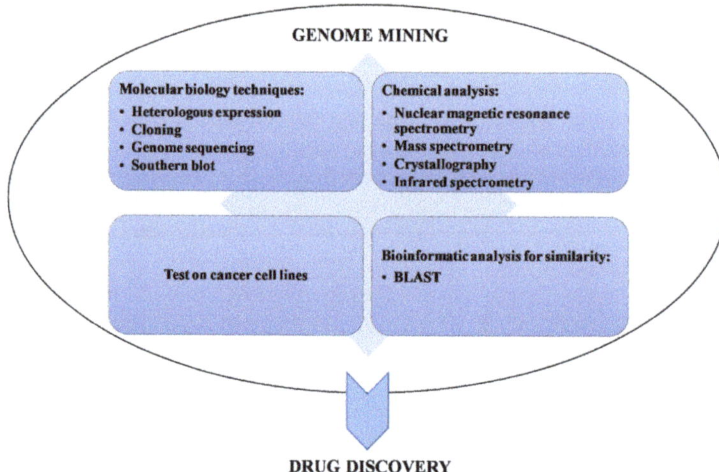

Figure 1. Associated techniques (categorized as molecular biology techniques, chemical analysis, cellular biology techniques, and bioinformatic analysis) to genome mining for validation of data, leading together to drug discovery.

Genome mining is entirely dependent on computing technology and bioinformatics tools. About this point, a huge amount of data, consisting of DNA sequences and their annotations, are now deposited in publicly accessible databases. The storage and handling of these resources relies on the continued development of computers and the networks. Once all the genes within a new genome are identified, they can be compared with those of known functions in the public databases. Both raw and annotated genomic data, as well as bioinformatics tools, for sequence comparisons are freely available through the different websites. It also important to keep in mind that it is now a mandatory publication prerequisite of most scientific journals that sequence data from research involving novel DNA sequences is deposited in a publicly accessible database.

In the case for which the sequences of many proteins, encoding for enzymes, involved in natural product biosynthesis are deposited in these databases, it is relatively easy to identify pathways in which they are involved by sequence comparisons. The availability of these synthesis enzymes and the pathways in which they operate, together with the sequence comparisons with genes from which they arise, can certainly be used to identify homologs, and potentially the pathways, in the new organism under analysis. However, it is important to consider that many enzymes are similar in sequences but follow chemical processes that are slightly different, leading to a different pathway or very different final end product.

Furthermore, genome mining has a strong support by synthetic biology, consisting in the design and the construction of new biological, as for examples enzymes, genetic circuits and/or the redesign of existing biological systems. These combined approaches are mainly used to detect novel natural products in bacteria and fungi probably because of operon organization of their synthesis genes [13], allowing the control of transcriptional levels and also the association of their potential metabolic function [17]. Moreover, the central role of genome mining consists in finding new biosynthetic gene clusters (BGCs). In fact, the BGCs encode for two class of enzymes, polyketide synthases (PKS) and non-ribosomal peptide synthases (NRPS), which are the two most important biosynthetic routes responsible for the formation of natural products [18]. This approach also provides the possibility to compare target gene clusters to known gene clusters useful for the prediction of their function and structure using different associated web database [5]. In fact, although the genome mining allowed to find and identify the gene clusters responsible for the production of natural product synthesis, in the last decade web tools and databases have been integrated to improve the performance of this approach [15]. This scientific progress has enabled the development of three important web tools: (i) "antibiotics and Secondary Metabolite Analysis SHell" (antiSMASH), its first version was issued in 2011 and it is a web server able to associate the gene clusters identification with a series of specific algorithms for compounds analysis [19]. Therefore, this approach performs the prediction of sequences and offers a more detailed analysis of identified gene clusters and consequently gives the predicted image of amino acid stereochemistry structure [5]. (ii) "PRediction Informatics for Secondary Metabolomes" (PRISM), open-web tool, consisting of a genomic prediction of secondary metabolomes. Using different algorithms that compare the new genetic information with 57 virtual enzymatic reactions (such as adenylation, acyltransferase, and acyl-adenylating), this approach provides the possibility of obtaining a correspondence between known natural drugs and possible new ones [20]. (iii) "Integrated Microbial Genomes Atlas of Biosynthetic gene Clusters" (IMG/ABC) [21], launched in 2015, is a large open web database of known predicted microbial BGCs able to associate BGCs with secondary metabolites (SMs) and analyze both BGCs and SMs. In this way, it offers the ability of finding similar function between BGCs present in database and BGCs to be identified [22].

Starting from these general considerations, in the present review we want to emphasize the significance of genome mining approach to identify new natural products, also underlining the possible advantages and disadvantages of this technique. Moreover, we debate the associated techniques that can be applied following to genome mining for validation of data. Finally, we review the literature describing all novel natural drugs found from bacteria, fungi and other living organisms by genome mining approach, focusing on the examples available in the literature of the last ten years.

2. The Significance Genome Mining in Drug Discovery

Approximately half of clinically approved drugs (including antibiotics) are represented by natural products and their derivatives. Recently, the development of new bioinformatics, genetics and analytic tools, has provided new strategies for the discovery of natural products of biotechnological interest known as "combinatorial biosynthesis approaches" [23,24]. These techniques, together with bioinformatic approaches, have shown that the ability of organisms (particularly microorganisms) to produce bioactive natural products has been underestimated [25]. These organisms have been deeper explored through the sequencing of their genome and the application of genome mining approaches [26]. Genome analysis has shed light the presence of numerous biosynthetic gene clusters that could be involved in the synthesis of other secondary metabolites defined cryptic or orphan for their unknown origin [25].

Genome mining aims at predicting the genes that encode for new natural compounds of biotechnological interest by using several bioinformatic approaches [21]. The importance of genome mining is based on the urgent need to discover new drug entities due to the increased incidence of severe diseases (such as cancer) and the reduced efficacy of existing drugs [27]. Furthermore, the biosynthetic gene clusters contain elements that can be used to increase the production of both natural and engineered products by promoting costs reduction and their commercial use [26].

2.1. Strengths and Weaknesses of Genome Mining

As in the case of all approaches, also genome mining has strengths and weaknesses, summarized in Table 1. One of the advantages of using genome mining is to foster the detection of a large amount of bioactive natural compounds [6]. In addition, genome mining approach is relatively cheap and easy to apply in laboratory, and it requires no particular skills and/or experience of the operators [28]. Combining genome mining with genetic engineering techniques will make it possible to achieve maximum diversity of natural products [29]. This bioinformatic approach allows to predict the chemical structure of bioactive natural products, but forecasts are often difficult to formulate [28,29].

Table 1. Strengths and weakness in the use of genome mining.

Strengths	Weaknesses
Easy to apply for experimental procedures in laboratory	Not to predict biotechnological potential of the natural compounds
Cheap and easy to apply in laboratory	Only known biosynthetic gene clusters
To predict chemical structures of bioactive natural products	Difficulty to formulate chemical structures
No particular skills and/or experience of the operators	Too new approach that needs to be deepened

As reported in Wohlleben et al. [6], a great disadvantage of genome mining is that only known biosynthetic gene clusters can be identified [29]. Moreover, with this approach, it is not possible to predict the biological activities of the natural products identified [26]. However, genome mining is still an evolving technique [29], in fact, scientists are trying to improve this bioinformatic tool in order to reduce the limits of this approach.

2.2. Synthetic Biology and Other Experimental Techniques Associated with Genome Mining

Synthetic biology progresses have been possible thanks to the very recent advent of DNA sequencing and synthesis in molecular biology field. The distinguishable element of synthetic biology respect to the other traditional molecular biology approaches is represented by its focus on the design and construction of components which are core for example of enzymes and metabolic pathways [30]. These genomic assessments joined to microbial diversity provide the fundamental natural libraries for further engineering.

In this review we have not focused our attention on synthetic biology because a great number of reviews on the most obvious and popular applications of synthetic biology methodologies have been published from 2008 to today [30–45].

Natural product production using engineered microorganisms represent the more important application of synthetic biology in the biotechnological field for natural products. The most important of commercialized examples are represented by two compounds produced by fermentation in genetically modified yeast: The semisynthetic malaria drug artemisinin and the first consumer-market synthetic biology product, "natural" vanillin [46,47]. These successful application of synthetic biology opened new perspective in the exploration of microbes as sources of high-value compounds on an industrial scale.

Genome mining is followed by the identification of cryptic pathways using several strategies, known as "combinatorial biosynthesis", which that can be used in order to create novel genetic combinations of structural biosynthetic genes. These methods consist of gene activation/inactivation and mutasynthesis approaches. Gene inactivation involves the creation of a mutant organism, in which the biosynthetic gene cluster becomes inactive, thus eliminating the production of metabolites. The comparison between mutant organisms can be made by high-performance liquid chromatography/mass spectrometry (HPLC-MS), revealing the natural product absent in the mutant organism [26]. Therefore, gene inactivation needs as evidence of cluster involvement in compound biosynthesis [24]. Secondary metabolites come from precursors of primary metabolism, and their over-production is related to an enhanced protein synthesis [48]. However, in some cases, there are genes that produce specific precursors not provided by the primary metabolism. These precursors are usually used as starting units for example to the production of polyketides synthases (PKS) or non-ribosomal peptide synthetases (NRPS), which in turn produce natural compounds. Inactivation of genes involved in the biosynthesis of these precursors leads to non-productive mutants that can be used for the biosynthesis of new compounds by mutasynthesis or mutational biosynthesis [23,24,26]. If some genes are silent, it would be impossible to produce and test the biological activity of the natural product. It is therefore necessary to apply the activation of silent pathways under the control of a constitutive promoter or inactivating repressors [28]. In the final stages of metabolites biosynthesis, several enzymes such as, transferase, oxygenase, oxidase, peroxidase, and reductase, play a key role for further modifications [26].

Examples of Other Experimental Techniques

A method to identify new natural products with biotechnological potential combines the research of coding genes for a specific compound with the detection of bacterial resistance. This approach, called target-directed genome mining, relies on the identification of gene clusters without knowing the molecules produced [49].

Another method to identify a natural product is the one strain/many compounds (OSMAC) approach. This method is based on the systematic alteration of culture media or cultivation parameters to force the expression of cryptic genes. In addition, any metabolism deregulation system can be used to improve the production of secondary metabolites, leading to the discovery of new bioactive compounds. Many of these approaches involve the treatment of known chemicals that modify the structure of chromatin or the use of small molecules that re-shape and regulate secondary metabolism by inhibiting the synthesis of fatty acids [50].

Another technique associated with genome mining is the in vitro reconstruction of biosynthetic pathways that produce natural products. This technique can be used to generate highly pure intermediates, limiting side reactions such as the formation of toxic compounds and reducing protein-protein interactions [51].

Taking into account this background, we review on the new natural drugs found from bacteria, fungi and other living organisms by genome mining approach. We analyzed organisms that derive not only from marine environment but also from the terrestrial ones, considering that the genome mining and other techniques associated with it are still at the beginning for the discovery of bioactive compounds from the sea.

2.3. Bacteria

The first point that must be underlined is that the most of medicinal products described above derive from bacteria [6] (see Table 2). In fact, the available literature on genome mining mainly concerns these microorganisms. Specifically, soil and marine bacteria, such as actinomycetes, produce the greatest part of natural drugs identified in the last thirty years [52]. The actinomycetes can be isolated from various habitats, such as soil, sea deposits, sponges, corals, mollusks, seagrasses, and mangroves [53].

Table 2. Genome mining approaches applied to microorganisms.

Microorganism	Experimental Purpose	Associated Techniques	References
Actinomycetes	Identification of strains capable to produce halogen enzymes.	PCR screening and NMR spectroscopy	[54]
Streptomyces aizunensis NRRL B-11277	Elucidation of new antibiotic ECO-02301 structure	HPLC, MIC	[55]
Streptomyces roseosporus	Anti-infective agent arylomycin and its BGCs	IMS, MS and SST	[56]
Streptomyces roseosporus	Identification of stenothricin and its BGCs	MS/MS spectra, antiSMASH, NMR, BioMAP, Cytological profiling	[57]
Streptomyces exfoliatus UC5319, *Streptomyces arenae* TU469 and *Streptomyces avermitilis*	Biosynthetic gene clusters involved in the synthesis of pentalenolactone	Cloning, MS/MS spectra, H-NMR spectroscopy	[58]
Streptomycetes sp. M10	To determine biosynthetic gene clusters involved in the synthesis of natural products	PRC screening, BLASTP, antiSMASH, Artemis Release 12.0, RT-PCR, MALDI-TOF	[59]
Streptomyces sp., *Streptomyces roche*, *Streptomyces lividans* SBT5	Streptothricin and borrelidin biosynthetic gene clusters	Heterologous expression, HPLC, LC-MS, LEXAS method, antiSMASH	[60]
Streptomyces sp. Tü 6176:	BGCs of nataxazole	antiSMASH 2.0 heterologous expression, gene inactivation, antibiotic disc diffusion assay, test on cancer cell lines	[61]
Strepmomyces argillaceus ATCC12956	Argimycin biosynthetic gene cluster	AntiSMASH, test on cancer cell lines	[62]
Streptomyces sp. CBMAI 2042	Valinomycin biosynthetic gene cluster	Test on pathogens, *in silico* analyses	[63]
Bacillus, *Streptomyces*, *Micronospora*, *Paenibacillus*, *Kocuria*, *Verricosispora*, *Staphylococcus*, *Micrococcus*	Influence of isolation location on secondary metabolite production	Test on cancer cell lines, MS, GNPS	[64]
Streptomyces sp. MA37, *Norcardia brasiliensis*, *Actinoplanes* sp. N902-109	Identification of Fluorinases	overexpression of gene, vitro activity assay and 19F NMR	[65]
Streptomyces sp. PKU-MA00045	Aromatic polyketides	1H-NMR and 13C-NMR spectra, genome sequencing, BLAST	[53]
Streptomyces sp. SM17	Identification of Surugamide A	NCBI BLASTN, antiSMASH, NMR	[10]
Pseudovibrio sp. POLY-S9	BGCs of symbiotic bacteria and gene involved in symbiontic relationship	genome sequencing, antiSMASH	[66]
Vibrio harveyi	BGCs of spongosine and potential secondary metabolites	MS/MS-based molecular networking, nitric oxide assay, MLSA and BLAST, genome sequencing and antiSMASH	[67]
Planctomyces	Metabolic properties of these bacteria	antiSMASH, MicroArray	[68]
Diverse prokaryotic species	New aldolase enzymes	LC–MS, cloning, FPLC, HTS	[69]
Pseudoalteromonas luteviolacea	Violacein biosynthetic pathway	LC-MS/MS, antiSMASH	[8]
Anabaena variabilis PCC 7937, *Anabaena* sp. PCC 7120, *Synechocystis* sp. PCC 6803 and *Synechococcus* sp. PCC 6301	MAA biosynthetic gene cluster	MAA induction with radiation UVR, MAA extraction, HPLC, BLAST	[70]
Hapalosiphon welwitschii UH strain IC-52-3, *Westiella intricate UH* strain HT-29-1 and *Fischerella* sp. CC 9431	Hapalosine biosynthetic pathway	PCR screening, antiSMASH, Geneious version 6.1.7	[71]

Hornung et al. [54] applied genome mining to identify strains capable of producing halogenase enzymes, where halogenations represent an important feature for the biological activity of a great number of different natural products. *Escherichia coli* DH5a and *E. coli* XL1 Blue were used to identify the complete halogenase gene sequence and to build primer-specific probes for these genes. Moreover, genomic DNA was isolated from 550 strains of actinomycetes available in strain collections. Using specific primer probes, it has been demonstrated that some actinomycetes are able to produce halogen enzymes.

Furthermore, nuclear magnetic resonance spectroscopy (NMR spectroscopy) was applied to understand the structure of these molecules, revealing that they were not exactly like those already known in literature. *Streptomyces*, a type of actinomycetes gram-positive bacteria, have also extensively been studied.

In fact, McAlpine et al. [55] used the genome mining approach to identify new antibiotic ECO-02301, with a potent antifungal activity, from *Streptomyces aizunensis* NRRL B-11277 bacteria. This compound was active against several strains of human pathogenic fungi (*Candida albicans* ATCC10231, *Candida glabrata* ATCC 90030, *Candida lusitaniae* ATCC 200953, *Candida tropicalis* ATCC 200955, *Candida krusei* LSPQ 0309, *Saccaromyces cerevisiae* ATCC 9763, *Aspergillus fumigatus* ATCC 204305, *Aspergillus flavus* ATCC 204304, *Cryptococcus neoformans* ATCC 32045). To obtain the expression of this gene after the grown of bacteria in flask, the proteins were extracted and analyzed by high performance liquid chromatography (HPLC) monitored by a diode array detector (DAD) that detects the absorption in UV region and positive/negative mass spectrometry (MS) ions.

The analysis of the genome of *S. aizunensis* NRRL B-11277 helped the prediction of the structure of this compound with sufficient accuracy so to represent a guide for its isolation.

Furthermore, an anti-infective agent, called arylomycin, and its BGCs by *Streptomyces roseosporus* strains, were described using imaging mass spectrometry (IMS) and MS guided by genome mining approaches [56]. Specifically, *S. roseosporus* was co-cultured with two pathogen strains, *Staphylococcus aureus* and *Staphylococcus epidermidis*, and its genome has been sequenced. It was so demonstrated that *S. roseosporus* produced daptomycin, an antibiotic molecule. Moreover, they spotted *S. roseosporus* in the center of *S. aureus* and *S. epidermidis* cultures and after 36 hours of incubation, using IMS and MS, aptomycin ions have been not observed, but a cluster corresponding to the potassium adduct was found. These results suggested that *S. roseosporus* was also able to produce three additional antibiotics. Furthermore, to identify the biosynthetic gene cluster of these molecules, a peptide-genomic mining approach was applied, which relied on the short sequence tag (SST) from tandem very spectrometric data. With this approach, in fact, they established that these three molecules were arylomycins.

In a similar study, Liu et al. [57] demonstrated that *S. roseosporus*, in addition to the non-ribosomal peptide synthetase-derived molecular families and their gene subnetworks, todaptomycin, arylomycin, and napsamycin, was also able to produce stenothricin. Firstly, after DNA extraction, to identify the molecular network they reduced the complexity of analysis to 837 genes using MS/MS spectra with parent ion masses within 0.3 Da and compared to related MS/MS spectral patterns. It was possible to observe the already known genes ofarylomycin, napsamycin, daptomycin and their variants. However, they identified four genes for stenothricin but combining the MS/MS spectra to the amino-acid blocks found by antiSMASH, 21 genes clusters were found. Furthermore, to understand their biological activities, a screening platform (named BioMAP) was used and then the cytological profiling, evaluating this activity against 15 bacterial strains. These approaches revealed that the stenothricinis was active on both Gram-negative and Gram-positive bacteria.

Seo et al. [58] used the DNA extraction to isolate the antibiotic pentalenolactone biosynthetic gene clusters from the known pentalenolactone producers *Streptomyces exfoliatus* UC5319 and *Streptomyces arenae* TU469. By building probes based on the previously cloned *S. exfoliates* pentalenene synthase gene, the sequence of the *S. exfoliatus* Pen biosynthetic gene cluster were analyzed, revealing that the furthest upstream gene, designated as PenR, encoded a 153 aa MarR-family transcriptional regulator. Moreover, PenI, PenH, and PenF were also found, which were expected to catalyze the oxidative conversion of pentalenene to 1-deoxy-11-oxopentalenic acid, as previously established for the othologous *Streptomyces avermitilis* proteins. Furthermore, the attention was pointed out on penE product, because it seems to be the key branch point that distinguished the pentalenolactone and neopentalenolactone biosynthetic pathways. PenE gene encoded a protein that is a homologue of the known Baeyer-Villiger monooxygenase from *S. avermitilis*, PtlE. The compounds PenD, PntD, and PtlD were characterized by mass spectrometry and H-NMR, also generating the deletion mutants with no production of pentalenolactone.

In another study, Tang et al. [49] analyzed, through bioinformatic approaches (BLASTP, Artemis Release 12.0), the genome of *Streptomycetes* sp. M10 discovering 20 biosynthetic gene clusters involved in the synthesis of natural products, such aspolyketides, NRPs, siderophores, lantibiotics, terpenoids. In addition, one of all gene clusters shared a partial similarity with candicidin/FR-008gene cluster, which in turn encoded for antifungal polyene assuming the potential role of this strain to produce this compound. Finally, to confirm this potential activity, the polyene was tested against the phytopathogen *Fulvia fulva* for its antifungal activity.

A high throughput genomic library expression analysis system (LEXAS) was applied for efficient, function-driven discovery of cryptic and new antibiotics from *Streptomyces*, known producers of several antibiotics [60]. Each BAC clone was transferred individually into an engineered antibiotic overproduction host, avoiding preference for smaller BACs. The LEXAS captured two known antibiotics, identified two novel lipopeptides and their BGC that was not produced/expressed in the native *Streptomyces rochei* strain, and revealed a cryptic BGC for unknown antibiotic. Specifically, in this research two new antibiotics streptothricins and borrelidin were found and for their validation these genes were expressed in the surrogate host *Streptomyces lividans* SBT5 by heterologous expression. Moreover, to analyze the antimicrobial activity, SBT5 products were tested against *Staphylococcus aureus* and *Bacillus mycoides*, showing an inhibition. In addition, they discovered two novel linear lipopeptides and their BGCs also adding the analysis of their structures by HPLC and liquid chromatography-mass spectrometry (LC-MS).

Thirty-eight secondary biosynthetic gene clusters of nataxazole (NAT) and its derivatives were identified from *Streptomyces* sp. Tü 6176, using in silico by genome mining and antiSMASH 2.0 [61]. In particular, the NAT entire BGC was described, consisting of 21 genes: 12 encoding for structural proteins, 4 for regulatory proteins, 4 probably involved in NAT secretion, and 1 with unknown function. Moreover, using the gene inactivation and heterologous expression of NAT cluster, it was established that secondary metabolite pathways were outside of NAT gene cluster (not a common in actinomycetes) despite they were involved in NAT biosynthesis. Furthermore, using antibiotic disc diffusion assay, an antibiotic activity was found only against *Staphylococcus albus* J1074, whereas the negative effect was absent in *Streptomyces lividans* JT46, *Micrococcus luteus* and *Escherichia coli*. Anticancer activity was tested against human tumor cell lines (HT29, A549, MDA-MB-231, AGS and A2780) including mouse cell line NIH/3T3 used as control. In this way, they demonstrated that these compounds have moderate activity against maleficent cells. In a similar study, Ye et al. (2017) used genome mining and antiSMASH 2.0 to identify the presence of 31 biosynthetic clusters in *Strepmomyces argillaceus* ATCC12956.

The most studied BGC between all found was the gene that encoded for argimycin P (renamed *arp* cluster). In addition, the pathway for the biosynthesis of *arp* was reconstructed by means of genetic engineering. Moreover, using in vitro tests on cells, no cytotoxic activity of this compound was found against 59 tumour cell lines. In another study, Paulo et al. [63] used in silico genome mining on strains of *Streptomyces* sp. CBMAI 2042, isolated from the branches of the plants *Citrus sinensis*. Moreover, this strain also prevented the proliferation of pathogens in citrus such as *Citrus xylella*, *Geotrichum candid* var. citri-Aurantii, and *Colletotrichum gloesporioides*. In particular, 35 biosynthetic gene clusters were found including the putative NRP biosynthetic gene cluster that encoded for valinomycin. In addition, by combining genome mining and molecular network, it was possible to reconstruct the origin of the biosynthetic pathway of cyclodepsipeptides, which have antibacterial, antiviral, and anticancer activity.

Furthermore, Purves et al. [64] applied the genome mining approach on bacteria extracted from two marine sediments (Antarctic and Scotland). They identified eight genera (*Bacillus*, *Streptomyces*, *Micromonospora*, *Paenibacillus*, *Kocuria*, *Verrucosispora*, *Staphylococcus*, and *Micrococcus*) and used 38 strains on which MS analysis was conducted. Thanks to this approach a great number of metabolites were identified, of which 1422 were Antarctic-specific, while 1501 were Scottish-specific secondary metabolites. Moreover, a molecular network was built up by Global Natural Products Social (GNPS) Molecular Networking, showing that only 8% of strains belonging to these locations displayed a similarity, implying a high degree of biogeographic influence upon secondary metabolite production. Organic extracts from these 38 selected strains were tested for cytotoxicity against epithelial colon adenocarcinoma cells (Caco-2) and human fibroblasts originating from foreskin (HS27). No effect on normal cell viability was observed, while seven extracts were bioactive against Caco-2 at 50 g/mL concentration. Direct observation revealed morphological changes, such as cell shrinkage and formation of apoptotic bodies. Moreover, Deng et al. [65] identified three new fluorinase enzymes from three bacterial strains, *Streptomyces* sp. MA37, *Nocardia brasiliensis*, and *Actinoplanes* sp. using the genome mining approach. These proteins were isolated and purified using overexpression of fluorinasegenes in *Escherichia coli*. Analyzing this product with in vitro activity assay, it revealed a high homology (about 85%) of its BGCs to the original one (called flA1) founded in *Streptomyces cattleya*. Finally, it was also assessed that *Streptomyces* sp. MA37produced some unidentified fluorometabolites.

As mentioned before, the actinomycetes are distributed in different marine habitats, being mainly associated to sponges. In fact, Jin et al. [53] have conducted genome mining experiments with *Streptomyces* sp. PKU-MA00045 isolated from sponges. Specifically, five new aromatic polyketides, fluostatins M–Q (1–5) were isolated using PCR-based genome mining method, and their chemical structures were clarified by ^{1}H-NMR and ^{13}C-NMR. The entire genome of *Streptomyces* sp. PKU-MA00045 was sequenced and compared to homologues in the published fluostatin gene clusters with BLAST, so identifying the BGCs of these new five compounds. In a similar experiment, Almeida et al. [10] used OSMAC approach to identify an octapeptidicsurugamide (Surugamide A) from *Streptomyces* sp. SM17, isolated from the marine sponge *Haliclona simulans*. The phylogenetic analysis with NCBI BLASTN demonstrated that this marine bacteria was phylogenetically linked to five strains of terrestrial *Streptomyces* bacteria: *Streptomyces albidoflavus* strain J1074, *Streptomyces albidoflavus* strain SM254, *Streptomyces sampsonii* strain KJ40, *Streptomyces* sp. FR-008 and *Streptomyces koyangensis* strain VK-A60T. Since *S. albidoflavus* strain J1074 was widely used as a model for various biotechnological studies, the secondary metabolites of the biosynthetic gene clusters were predicted by antiSMASH program, comparing the new BGCs with those already collected by *S. albidoflavus* strains. In this way, it was demonstrated that *Streptomyces* sp. SM17 produced different secondary metabolites. Moreover, using NMR technique it was possible to show that *Streptomyces* sp. SM17 was able to produce higher levels of Surugamide A than the *S. albidoflavus* strain J1074.

However, Anoop et al. [66] studied another bacterial strain *Pseudovibrio* sp. POLY-S9, isolated from intertidal marine sponge *Polymastia penicillus* sampled from the Atlantic coast of Portugal. In fact, after genome sequencing of this marine bacteria, new genes-related bioactive compounds were isolated, such as polyketide synthase, nonribosomal peptide synthetase and siderophore, using genome mining by antiSMASH. Moreover, several genes involved in symbiotic relationships, such as the ankyrin repeats, tetratrico peptide repeats and Sel1, were also identified. Another important finding of this study was represented by some genome plasticity elements of POLYS9, which allowed the survival of these bacteria and their adaptation to various habitats through the exchange of genetic material. Using MS/MS-based molecular networking analysis a bacterial strain was isolated from the Caribbean sponge *Tectitethya crypta*, able to produce spongosine, deoxyspongosine, spongothymidine, and spongouridine, generally referred as "spongonucleosides" [67].

Spongosine, a methoxyadenosine derivative, had several pharmacological applications, having anti-inflammatory activity (for their capability to inhibit the nitric oxide production in cells) and analgesic and vasodilation properties. After MLSA and BLAST analyses, this strain was identified as *Vibrio harveyi*, and thanks the genomic DNA sequencing and antiSMASH platform, six potential secondary metabolite pathways were described.

Planctomycetes are ubiquitous bacteria that were usually found in marine, freshwater and soil habitats, even if it is possible to find them as free living organisms, or attached to abiotic and biotic surfaces, as for example to algal cells. Some strains also live as symbionts of prawns, marine sponges or termites [72]. For instance, Jeske et al. [68] applied the genome mining methods to define the metabolic properties of *Planctomyces*. First, they found 102 genes or gene clusters involved in the production of secondary metabolites by analyzing 13 genomes on antiSMASH database. Moreover, the genome analysis showed a close correlation between the length of BGCs and the amino acid sequence of the predicted secondary metabolites. Moreover, since most BGCs were transcriptional silent, the Phenotype MicroArray technology was applied on compounds secreted by *Planctomyces limnophilus* (limnic strain) and *Rhodopirellula baltica* (marine strain), confirming that there was a strong relationship between *Planctomycetes* and algae or plants, which in turn secrete compounds that might serve as trigger to stimulate the secondary metabolite production in *Planctomycetes*. Thus, this study provides strong evidences for the use of these bacteria for drug development.

In a different study, Guérard-Hélaine et al. [69] identified new aldolase enzymes, belonging the aldolase/transaldolase family, from 313 different prokaryote species. Comparing the sequence of 1148 proteins extracted from these strains to already known aldolases and transaldolases, 700 genes were selected. The overexpression of these genes and the following LC-MS analysis allowed the selection of 19 proteins of interest. After cloning of the corresponding genes and using fast protein liquid chromatography (FPLC), 18 enzymes were purified, including two aldolases and sixteen transaldolase. Moreover, the activity of these 18 enzymes was evaluated by high-throughput screening (HTS), revealing that six of those annotated as transaldolase showed aldolase activity. Maansson et al. [8] extracted DNA from 13 closely related strains identified as *Pseudoalteromonas luteviolacea*, isolated from all over the Earth, and analysed their potential to produce secondary metabolites. Specifically, antiSMASH analysis demonstrated that only 10 biosynthetic pathways were preserved in all strains, including glycosylated lantipeptide (RiPP1) and two bacteriocins (RiPP2 and RiPP3). All strains have maintained essential pathways, such as that responsible for the production of siderophores, homoserine lactones and violacein. Furthermore, bacteria were grown in culture media to stimulate the synthesis of secondary metabolites and the chemical structures of these compounds were analyzed by LC-MS/MS. Particular attention was paid on violacein pathway, showing the presence of an insert in the *bmp1* gene in the thioesterase domain probably responsible of *Pseuoalteromonas* color. Moreover, the varieties *Pseudoalteromonas* S4047-1, S4054 and CPMOR-1 produced indolmicin antibiotic. However, the biosynthetic pathway coding for the antibiotic indolmicin has never been characterized.

Cyanobacteria

Cyanobacteria were also studied for their interesting bioactive secondary metabolites. For example, they produce mycosporine and mycosporine-like amino acids (MAAs), which are antioxidant molecules that eliminate toxic oxygen radicals protecting cells from saline, drying or thermal stress in some organisms and may act as an intracellular nitrogen reservoir. These compounds were also found in many other organisms such as yeasts, fungi, algae, corals and lichens [73]. Applying genome mining approach and BLAST analysis, Singh et al. [70] demonstrated that among four strains of cyanobacteria (*Anabaena variabilis* PCC 7937, *Anabaena* sp. PCC 7120, *Synechocystis* sp. PCC 6803 and *Synechococcus* sp. PCC 6301) exposed to 72 hours of UV radiation, only *Anabaena variabilis* PCC 7937 was able to produce MAAs. HPLC analysis of these four cyanobacteria revealed the presence of a unique combination of two genes, predicted *DHQ synthase* (YP\324358) and *O-methyltransferase*; (YP\324357) in *A. variabilis* PCC7937, which were missing in other non-MAAs-synthesizing cyanobacteria. Micallef et al. [71] identified the gene cluster responsible for hapalosine synthesis and hapalosine biosynthetic pathway from the genomes of three cyanobacteria (*Hapalosiphon welwitschii* UH strain IC-52-3, *Westiella intricata* UH strain HT-29-1 and *Fischerella* sp. CC 9431), by using genome mining combined with

Geneious version 6.1.7 and antiSMASH. Single cyanobactin cluster of biosynthetic genes was identified only in the genome of *W. intricate* UH strain HT-29-1, demonstrating that there is structural diversity of cyanobacteria inside cyanobacteria strains. Moreover, only *Fischerella* sp. PCC 9339 encoded a microviridine gene cluster and they identified the MAA (*mys*) gene cluster in the strains *W. intricate* varieties UH HT-29-1, *H. welwitschii* UH strain IC-52-3, *Mastigocoleus testarum* BC008, *Fischerella muscicola* SAG1427–1 and *Chloroglopsis* sp. PCC 9212. Finally, the presence of the cluster of scytonemin genes within the genome of *Mastigocladopsis repens* PCC 10,914 was discovered, suggesting that this organism was able to bio-sintetizes cytonemin in order to protect the cells against UVA-radiation. The geosmin gene cluster was identified in *W. intricata* variety UH HT-29-1, *H. welwitschii* UH strain IC-52-3, *Fischerella* sp. PCC 9431, and *F. muscicola* SAG 1427–1.

2.4. Fungi

As described above, the most important sources of natural drugs are not only bacteria but also fungi [6]. In fact, many different natural products, such as penicillin, cephalosporin, ergotrate and the statins represent well-known fungal secondary metabolites for pharmacological applications [74]. For these organisms the genome mining also proved to be a useful method to find BGCs (Table 3). In a study of Bergmann et al. [75] a silent metabolic pathway was detected, which might code for the biosynthesis of polyketides or polypeptides in *Aspergillus nidulans*. In particular, considering that the cryptic gene cluster provided a putative activator gene called *apdR*, it was amplified and cloned into expression vector pAL4, which coded for inducible alcohol dehydrogenase promoter *alcAp* of *A. nidulans* and the pyr-4 gene of *Neurospora crassa* as a selectable marker.

Table 3. Genome mining approaches applied to fungi.

Fungi	Experimental Purpose	Associated Techniques	References
Aspergillus nidulans	Detection of silent metabolic pathway	Southern blot, HPLC, NMR, IR, and MS	[75]
Calcarisporium arbuscula	Silent metabolic pathway involved in natural product biosynthesis	genome sequencing, LC-MS, chromatographic and NMR analysis, HPLC	[76]
Aspergillus MF297-2	Identification of BGCs of ephacidin and notoamide	genome sequencing, BLAST, gene cloning, overexpression of protein, HPLC, LC-MS, 1H, and 13C NMR	[77]
Aspergillus oryzae and *Neosartorya fischeri*	Isolation of terpene synthases	heterologous expression, GC-MS, 1H- and 13C-NMR, LC-MS, and HR-MS	[78]

Using Southern blot analysis, it was demonstrated that under inducing conditions the *apdA* gene encoded the PKS-NRPS hybrid synthetase. Moreover, HPLC analysis displayed that this induced strains were able to produce two main products, Aspyridones A and B, and two minor compounds, whose structures was elucidated by NMR and MS. In a similar study, Mao et al. [76] revealed a silent metabolic pathway involved in natural product biosynthesis. In fact, after genome sequencing, 68 BGCs were identified, being in contrast to the two predominant metabolites normally produced, the F1-ATPase inhibitors 1 and 2. Since these BGCs are localized within the heterochromatic regions, a mutant strain was built deleting *hdaA* (gene of the histone H3 lysine 14 (K14) deacetylase). In this way, using metabolite extraction and LC-MS analysis, it was demonstrated that the mutant produced more compounds compared to wild strain. Moreover, after overexpression of these genes, ten compounds were isolated, of which four contained new structures, including the cyclic peptides arbumycin and arbumelin, the diterpenoid arbuscullic acid A, and the meroterpenoid arbuscullic acid B. However, Ye et al. [78] applied the genome mining approach to conduct a phylogenetic analysis of fifteen bifunctional terpene synthases found in five fungal genomes. Specifically, the terpene BGCs sequence were different and synthetized sesterterpenes with new carbon skeletons, suggesting that these microorganisms were separated in five different clades. Moreover, two clades, *Aspergillus oryzae* and *Neosartorya fischeri*, did not produce terpene, hypothesizing that BGCs were silent in standard conditions. For these reasons, heterologous expression was performed in *A. oryzae* using *E. coli* plasmids and the extract was analyzed with GC-MS, ^{1}H- and ^{13}C-NMR elucidating the structure of four compounds, one of which known as sesterfisherolsynthase (*NfSS*) and previously found in *N. fischeri*. Furthermore, bioinformatic analysis showed that *NfSS* gene was encoded downstream of a cytochrome P450 monooxygenase (NfP450) and it was transformed by NfP450 to sesterfisheric acid. Finally, to identify NfP450 gene, double transformant with NfSS and NfP450 genes was prepared and the extract was examined by LC-MS and HR-MS indicating that NfP450 conducted a NfSS modification.

Furthermore, Ding et al. [77] have identified the first BGCs of the stephacidin and notoamide, belong to family of prenylated alkaloids, from *Aspergillus* sp. MF297-2. Specifically, after sequencing of genome the entirenotoamide and stephacidin gene cluster was identified by BLAST comparing sequence to gene *ftmA*, which was previously mined from an *Aspergillus fumigatus*. By bioinformatics analysis, 19 genes involved in notoamide biosynthesis were found to constitute this cluster. To understand the function, this cluster was cloned using *E. coli* DH5R and overexpressed into *E. coli* BL21. The proteins were purified with a single Ni-NTA column and analyzed with HPLC, LC-MS, ^{1}H- and ^{13}C-NMR. Two central pathway enzymes, *NotF* and *NotC*, were identified suggesting a scheme for the biosynthesis of stephacidin and notoamide metabolites.

2.5. Other Organisms

Several other organisms, completely unrelated to the marine environment, have been used as subject of genome mining approach, such as terrestrial microorganisms, plants, and animals (Liu et al. 2018; Table 4).

Table 4. Genome mining approaches applied to ants, copepods and plants.

Organism	Experimental Purpose	Associated Techniques	References
Atta cephalotes, Camponotus floridanus and *Harpegnathos saltator*	Defense- and neuropeptides in Social Ants	tBLASTn, GeneWise algorithm, ClustalW	[4]
Calanus sp., *Pontella* sp., *Oikopleura* sp., *Acartia* sp., *Acartia* sp. and *Corycaeus* sp.	Metabolic pathway from conversion from β-carotene to astaxanthin.	LC-UV method, HPLC, Hhpred database	[79]
Arabidopsis thaliana, Capsella rubella, Brassica oleracea, Nicotiana benthamiana, Agrobacterium tumefaciens	Sesterterpene biosynthetic gene cluster	plantiSMASH, heteroloug expression, GC-MS, cristallography	[80]

Gruber and Muttenthaler [4] applied genome mining to identify defense- and neuropeptides in the genomes of social ants of the subfamilies of Myrmicinae (*Atta cephalotes*), Formicinae (*Camponotus floridanus*) and Ponerina (*Harpegnathos saltator*); ants are difficult to manipulate for scientific purposes because of the size of their bodies and organs. Most interestingly, genes encoding for oxytocin/vasopressin-related peptides (inotocins) and their putative receptors were identified, using a publicly available matrix of tools, including the search for similarity with tBLASTn, prediction of gene structure using GeneWise algorithm and alignments of sequences by ClustalW.

Carotenoids cannot be synthesized de novo, but they must therefore be taken with food (such as algae) and get protective human health benefits as well. Free astaxanthin and its esterified forms are the main carotenoids present in crustaceans and in particular in copepods. Mojib et al. [79] aimed on understanding the metabolic and genetic basis of the blue phenotype between the blue pigmented organisms from the phylum Arthropoda, subclass Copepoda (*Acartia fossae*) and the phylum Chordata, class Appendicularia (*Oikopleura dioica*) in the Red Sea. Firstly, liquid chromatography-UV method was used to detect the carotenoids and mass spectrometry and HPLC were used to detect intermediate metabolites, present at low concentrations. The chromatograms identified astaxanthin in all samples, while the fucoxanthin was not detected in any samples. In addition, other carotenoids, intermediate compounds for conversion from β-carotene to astaxanthin, were also identified. The metabolic pathway for each sample was reconstructed for the conversion from β-carotene to astaxanthin. The results showed that all the species followed the same metabolic pathways via almost the same intermediate metabolite formation. Echinenone, one of the intermediate metabolite was not detected in any of the samples but its hydroxylated form, the 3-idrossi chinenone, was detected in all samples, as well as lutein. Putative β-carotene hydroxylase of P450 family coding transcripts was identified in blue *A. fossae* by in silico transcriptome mining. Putative carotenoid-binding proteins after transcriptome/genome mining showing 100% homology to Apolipoprotein D (ApoD) and crustacyanin as predicted by HHpred database.

A customized version of the plantiSMASH genome mining algorithm was created to identify a sesterterpene synthase gene repertoire in some *Brassicaceae* plants, which synthesizes fungal-type sesterterpenes with diverse scaffolds, thus fueling the drug-discovery pipeline [80]. Sesterterpenoids are a rare terpene class with not well explored chemical structure and diversity, representing a potential new drug source. This study offered new insights on the origin of structural diversity for protein engineering, supporting the idea of convergent evolution for natural product biosynthesis.

3. General Conclusions

Many drugs used, for example, as anticancer, antibacterial, and anti-inflammatory agents in the clinical field are derived from natural products and their derivatives. In fact, these secondary metabolites are produced by all organisms (from bacteria to plants, invertebrates, and other animals) and show several biological activities useful in several biotechnological applications. However, the most important sources of natural drugs are microorganisms (mainly bacteria, also associated with marine organisms, such to sponges) and fungi. In the last decades, the great advances made in the field of molecular biology techniques, representing a good example the genome mining together with the synthetic biology, strongly push the identification of BGCs, encoding for enzymes involved in the biosynthesis of natural products. Taking together, these next-generation and highly sophisticated tools contribute to the emergence of a new generation of natural product research. These techniques are in their infancy for their application to marine environment, but there are in literature a lot of applications for the discovery of bioactive natural products for other environments. For this reason, we think that a review reporting all these examples could give strong support in pushing the applications of these new techniques in discovery bioactive compounds from the marine environment, also due to high level of biodiversity offered by the sea in comparison with the Earth. Genome mining, as well as synthetic biology and all the techniques to them associated, represent a new challenge in natural products discovery from the marine environment, without impact on the environment and with no use of destructive collection practices of marine organisms.

Author Contributions: M.C. conceived and designed the scheme of the review; L.A. and R.E. performed the bibliographic research and prepared the original draft; N.R. contributed to the preparation of the original draft; All authors have read and agree to the final version of the manuscript.

Funding: This research received no external funding.

Acknowledgments: Luisa Albarano was supported by a PhD (PhD in Biology, University of Naples Federico II) fellowship co-funded by the Stazione Zoologica Anton Dohrn and University of Naples Federico II. Roberta Esposito was supported by a PhD (PhD in Biology, University of Naples Federico II) fellowship funded by the Photosynthesis 2.0 project of the Stazione Zoologica Anton Dohrn. Nadia Ruocco was supported by a research grant "Antitumor Drugs and Vaccines from the Sea (ADViSE)" project (PG/2018/0494374).

Conflicts of Interest: The authors declare no conflict of interest.

References

1. Ren, H.; Wang, B.; Zhao, H. Breaking the silence: New strategies for discovering novel natural products. *Curr. Opin. Biotechnol.* **2017**, *48*, 21–27. [CrossRef]
2. Newman, D.J.; Cragg, G.M. Marine-sourced anti-cancer and cancer pain control agents in clinical and late preclinical development. *Mar. Drugs* **2014**, 255–278. [CrossRef]
3. Gruber, C.W. Global cyclotide adventure: A journey dedicated to the discovery of circular peptides from flowering plants. *Pept. Sci.* **2010**, *94*, 565–572. [CrossRef]
4. Gruber, C.W.; Muttenthaler, M. Discovery of defense- and neuropeptides in social ants by genome-mining. *PLoS ONE* **2012**, *7*, 1–12. [CrossRef]
5. Boddy, C.N. Bioinformatics tools for genome mining of polyketide and non-ribosomal peptides. *J. Ind. Microbiol. Biotechnol.* **2014**, 443–450. [CrossRef]
6. Wohlleben, W.; Mast, Y.; Stegmann, E.; Ziemert, N. Antibiotic drug discovery. *Microb. Biotechnol.* **2016**, *9*, 541–548. [CrossRef]
7. Peláez, F. The historical delivery of antibiotics from microbial natural products—Can history repeat? *Biochem. Pharmacol.* **2006**, *71*, 981–990. [CrossRef]
8. Maansson, M.; Vynne, N.G.; Klitgaard, A.; Nybo, J.L.; Melchiorsen, J.; Nguyen, D.D.; Sanchez, L.M.; Ziemert, N.; Dorrestein, P.C.; Andersen, M.R.; et al. An integrated metabolomic and genomic mining workflow to uncover the biosynthetic potential of bacteria. *Am. Soc. Microbiol.* **2016**, *1*, 1–14. [CrossRef]
9. Gulder, T.A.M.; Moore, B.S. Chasing the treasures of the sea-bacterial marine natural products. *Curr. Opin. Microbiol.* **2009**, *12*, 252–260. [CrossRef]
10. Almeida, E.L.; Kaur, N.; Jennings, L.K.; Felipe, A.; Rinc, C.; Jackson, S.A.; Thomas, O.P.; Dobson, A.D.W. Genome mining coupled with OSMAC-based cultivation reveal differential production of Surugamide A by the marine sponge isolate *Streptomyces* sp. SM17 when compared to its terrestrial relative *S. albidoflavus* J1074. *Microorganisms* **2019**, *7*, 394. [CrossRef]
11. Tracanna, V.; De Jong, A.; Medema, M.H.; Kuipers, O.P. Mining prokaryotes for antimicrobial compounds: From diversity to function. *FEMS Microbiol. Rev.* **2017**, *41*, 417–429. [CrossRef]
12. Durand, G.A.; Raoult, D.; Dubourg, G. Antibiotic discovery: History, methods and perspectives. *Int. J. Antimicrob. Agents* **2019**, *53*, 371–382. [CrossRef]
13. Nett, M. Genome mining: Concept and strategies for natural product discovery. *Prog. Chem. Org. Nat. Prod.* **2014**, *99*, 199–245.
14. Challis, G.L. Genome mining for novel natural product discovery. *J. Med. Chem.* **2008**, *51*, 2618–2628. [CrossRef]
15. Ziemert, N.; Alanjary, M.; Weber, T. The evolution of genome mining in microbes—A review. *Nat. Prod. Rep.* **2016**, *33*, 988–1005. [CrossRef]
16. Trivella, D.B.B.; De Felicio, R. The tripod forbacterial natural product discovery: Genome mining, silent pathway induction, and mass spectrometry-based molecular networking. *mSystems* **2018**, *3*, e00160–e17. [CrossRef]
17. Koonin, E.V. Evolution of genome architecture. *Int. J. Biochem. Cell Biol.* **2009**, *41*, 298–306. [CrossRef]
18. Timmermans, M.L.; Paudel, Y.P.; Ross, A.C. Investigating the biosynthesis of natural products from marine proteobacteria: A survey of molecules and strategies. *Mar. Drugs* **2017**, *15*, 235. [CrossRef]
19. Blin, K.; Medema, M.H.; Kazempour, D.; Fischbach, M.A.; Breitling, R.; Takano, E.; Weber, T. antiSMASH 2.0—A versatile platform for genome mining of secondary metabolite producers. *Nucleic Acids Res.* **2013**, *41*, 204–212. [CrossRef]
20. Skinnider, M.A.; Dejong, C.A.; Rees, P.N.; Johnston, C.W.; Li, H.; Webster, A.L.H.; Wyatt, M.A.; Magarvey, N.A. Genomes to natural products PRediction Informatics for Secondary Metabolomes (PRISM). *Nucleic Acids Res.* **2015**, *43*, 9645–9662. [CrossRef]
21. Machado, H.; Tuttle, R.N.; Jensen, P.R. Omics-based natural product discovery and the lexicon of genome mining. *Curr. Opin. Microbiol.* **2017**, *39*, 136–142. [CrossRef]
22. Hadjithomas, M.; Chen, I.A.; Chu, K.; Huang, J.; Ratner, A.; Palaniappan, K.; Andersen, E.; Markowitz, V.; Kyrpides, N.C.; Ivanova, N. IMG-ABC: New features for bacterial secondary metabolism analysis and targeted biosynthetic gene cluster discovery in thousands of microbial genomes. *Nucleic Acids Res.* **2017**, *45*, 560–565. [CrossRef]

23. Helfrich, E.J.N.; Reite, S.; Piel, J. Recent advances in genome-based polyketide discovery. *Curr. Opin. Biotechnol.* **2014**, *29*, 107–115. [CrossRef]

24. Olano, C.; Méndez, C.; Salas, J.A. Molecular insights on the biosynthesis of antitumour compounds by actinomycetes. *Microb. Biotechnol.* **2011**, *4*, 144–164. [CrossRef]

25. Nett, M.; Ikeda, H.; Moore, B.S. Genomic basis for natural product biosynthetic diversity in the actinomycetes. *Nat. Prod. Rep.* **2009**, *26*, 1362–1384. [CrossRef]

26. Olano, C.; Méndez, C.; Salas, J.A. Strategies for the design and discovery of novel antibiotics using genetic engineering and genome mining. In *Antimicrobial Compounds: Current Strategies and New Alternatives*; Villa, T.G., Veiga-Crespo, P., Eds.; Springer: Berlin/Heidelberg, Germany, 2014; pp. 1–25.

27. Mody, K.H.; Haldar, S. Genome mining for bioactive compounds. In *Springer Handbook of Marine Biotechnology*; Kim, S.K., Ed.; Springer Handbooks; Springer: Berlin/Heidelberg, Germany, 2015; pp. 531–539.

28. Scheffler, R.J.; Colmer, S.; Tynan, H.; Demain, A.L.; Gullo, V.P. Antimicrobials, drug discovery, and genome mining. *Appl. Microbiol. Biotechnol.* **2013**, *97*, 969–978. [CrossRef]

29. Zerikly, M.; Challis, G.L. Strategies for the discovery of new natural products by genome mining. *ChemBioChem* **2009**, *10*, 625–633. [CrossRef]

30. Breitling, R.; Takano, E. Synthetic biology advances for pharmaceutical production. *Curr. Opin. Biotechnol.* **2015**, *35*, 46–51. [CrossRef]

31. Keasling, J.D. Synthetic biology for dynthetic chemistry. *ACS Chem. Biol.* **2008**, *3*, 64–76. [CrossRef]

32. Quin, M.B.; Schmidt-Dannert, C. Designer microbes for biosynthesis. *Curr. Opin. Biotechnol.* **2014**, *29*, 55–61. [CrossRef]

33. Singh, V. Recent advancements in synthetic biology: Current status and challenges. *Gene* **2014**, *535*, 1–11. [CrossRef]

34. Sleator, R.D. The synthetic biology future. *Bioengineered* **2014**, *5*, 69–72. [CrossRef]

35. Unkles, S.E.; Valiante, V.; Mattern, D.J.; Brakhage, A.A. Synthetic biology tools for bioprospecting of natural products in eukaryotes. *Chem. Biol.* **2014**, *21*, 1–7. [CrossRef]

36. Wilson, M.C.; Piel, J. Metagenomic approaches for exploiting uncultivated bacteria as a resource for novel biosynthetic enzymology. *Chem. Biol.* **2013**, *20*, 636–647. [CrossRef]

37. Wright, G. Synthetic biology revives antibiotics. *Nature* **2014**, *509*, S13.

38. Medema, M.H.; Breitling, R.; Bovenberg, R.; Takano, E. production in microorganisms. *Nat. Publ. Gr.* **2010**, *9*, 131–137.

39. Cobb, R.E.; Luo, Y.; Freestone, T.; Zhao, H. Drug discovery and development via synthetic biology. In *Synthetic Biology—Tools and Applications*; Zhao, H., Ed.; Elsevier: Oxford, UK, 2013; pp. 183–206.

40. Hranueli, D.; Starcevic, A.; Zucko, J.; Rojas, J.D.; Diminic, J.; Baranasic, D.; Gacesa, R.; Padilla, G.; Long, P.F.; Cullum, J. Synthetic biology: A novel approach for the construction of industrial microorganisms. *Food Technol. Biotechnol.* **2013**, *51*, 3–11.

41. Zakeri, B.; Lu, T.K. Synthetic biology of antimicrobial discovery. *ACS Synth. Biol.* **2012**. [CrossRef]

42. Cummings, M.; Breitling, R.; Takano, E. Steps towards the synthetic biology of polyketide biosynthesis. *FEMS Microbiol. Lett.* **2014**, *351*, 116–125. [CrossRef]

43. Genilloud, O. The re-emerging role of microbial natural products in antibiotic discovery. *Antonie Van Leeuwenhoek* **2014**, *106*, 173–188. [CrossRef]

44. Luo, Y.; Cobb, R.E.; Zhao, H. Recent advances in natural product discovery. *Curr. Opin. Biotechnol.* **2014**, *30*, 230–237. [CrossRef]

45. Porro, D.; Branduardi, P.; Sauer, M.; Mattanovich, D. Old obstacles and new horizons for microbial chemical production. *Curr. Opin. Biotechnol.* **2014**, *30*, 101–106. [CrossRef]

46. Paddon, C.J.; Keasling, J.D. Semi-synthetic artemisinin: A model for the use of synthetic biology in pharmaceutical development. *Nat. Rev. Microbiol.* **2014**, *12*, 355–367. [CrossRef]

47. Kurita, K.L.; Glassey, E.; Linington, R.G. Integration of high-content screening and untargeted metabolomics for comprehensive functional annotation of natural product libraries. *Proc. Natl. Acad. Sci. USA* **2015**, *112*, 11999–12004. [CrossRef]

48. Wang, G.; Hosaka, T.; Ochi, K.; Wang, G.; Hosaka, T.; Ochi, K. Dramatic activation of antibiotic production in *Streptomyces coelicolor* by cumulative drug resistance mutations. *Appl. Environ. Microbiol.* **2008**, *74*, 2834. [CrossRef]

49. Tang, X.; Li, J.; Milla, N.; Zhang, J.J.; Neill, E.C.O.; Ugalde, J.A.; Jensen, P.R.; Mantovani, S.M.; Moore, B.S. Identification of thiotetronic acid antibiotic biosynthetic pathways by target-directed genome mining. *ACS Chem. Biol.* **2015**, *10*, 2841–2849. [CrossRef]

50. Craney, A.; Ozimok, C.; Pimentel-elardo, S.M.; Capretta, A.; Nodwell, J.R. Chemical perturbation of secondary metabolism demonstrates important links to primary metabolism. *Chem. Biol.* **2012**, *19*, 1020–1027. [CrossRef]

51. Kwon, T.; Lee, G.; Rhee, Y.; Park, H.; Chang, M.; Lee, S.; Lee, J.; Lee, T. Identification of nickel response genes in abnormal early developments of sea urchin by differential display polymerase chain reaction. *Ecotoxicol. Environ. Saf.* **2012**, 1–7.

52. Newman, D.J.; Cragg, G.M. Natural products as sources of new drugs from 1981 to 2014. *J. Nat. Prod.* **2016**, *79*, 629–661. [CrossRef]

53. Jin, J.; Yang, X.; Liu, T.; Xiao, H.; Wang, G.; Zhou, M.; Liu, F.; Zhang, Y.; Liu, D.; Chen, M.; et al. Fluostatins M–Q featuring a 6-5-6-6 ring skeleton and high oxidized A-rings from marine *Streptomyces* sp. PKU-MA00045. *Mar. Drugs* **2018**, *16*, 87. [CrossRef]

54. Hornung, A.; Bertazzo, M.; Dziarnowski, A.; Schneider, K.; Welzel, K.; Wohlert, S.; Holzenkämpfer, M.; Nicholson, G.J.; Bechthold, A.; Süssmuth, R.D.; et al. A genomic screening approach to the structure-guided identification of drug candidates from natural sources. *ChemBioChem* **2007**, *8*, 757–766. [CrossRef]

55. Mcalpine, J.B.; Bachmann, B.O.; Piraee, M.; Tremblay, S.; Alarco, A.; Zazopoulos, E.; Farnet, C.M. Microbial genomics as a guide to drug discovery and structural elucidation: ECO-02301, a novel antifungal agent, as an example. *J. Nat. Prod.* **2005**, *68*, 493–496. [CrossRef]

56. Liu, W.; Kersten, R.D.; Yang, Y.; Moore, B.S.; Dorrestein, P.C. Imaging mass spectrometry and genome mining via short sequence tagging identified the anti-infective agent arylomycin in Streptomyces roseosporus. *J. Am. Chem. Soc.* **2011**, *133*, 18010–18013. [CrossRef]

57. Liu, W.; Lamsa, A.; Wong, W.R.; Boudreau, P.D.; Kersten, R.; Peng, Y.; Moree, W.J.; Duggan, B.M.; Moore, B.S.; Gerwick, W.H.; et al. MS/MS-based networking and peptidogenomics guided genome mining revealed the stenothricin gene cluster in *Streptomyces roseosporus*. *J. Antibiot. Tokyo* **2014**, *67*, 99–104. [CrossRef]

58. Seo, M.; Zhu, D.; Endo, S.; Ikeda, H.; Cane, D.E. Genome mining in *Streptomyces*. Elucidation of the role of baeyer-villiger monooxygenases and non-heme iron-dependent dehydrogenase/oxygenases in the final steps of the biosynthesis of pentalenolactone and neopentalenolactone. *Biochemistry* **2011**, *50*, 1739–1754. [CrossRef]

59. Tang, J.; Liu, X.; Peng, J.; Tang, Y.; Zhang, Y. Genome sequence and genome mining of a marine-derived antifungal bacterium *Streptomyces* sp. M10. *Appl. Microbiol. Biotechnol.* **2015**, *99*, 2763–2772. [CrossRef]

60. Xu, M.; Wang, Y.; Zhao, Z.; Gao, G.; Huang, S.-X.; Kang, Q.; He, X.; Lin, S.; Pang, X.; Deng, Z.; et al. Functional genome mining for metabolites encoded by large gene clusters using heterologous expression of a whole genomic BAC library in *Streptomyces*. *Appl. Environ. Microbiol.* **2016**, *82*, 5795–5805. [CrossRef]

61. Cano-prieto, C.; García-Salcedo, R.; Sánchez-Hidalgo, M.; Braña, A.F.; Fiedler, H.-P.; Méndez, C.; Salas, J.A.; Olano, C. Genome mining of *Streptomyces* sp. Tü 6176: Characterization of nataxazole biosynthesis pathway. *ChemBioChem* **2015**, *16*, 1461–1473. [CrossRef]

62. Ye, S.; Molloy, B.; Braña, A.F.; Zabala, D.; Olano, C.; Cortés, J.; Morís, F.; Salas, J.A.; Méndez, C. Identification by genome mining of a Type I Polyketide gene cluster from *Streptomyces argillaceus* involved in the biosynthesis of pyridine and piperidine alkaloids Argimycins P. *Front. Microbiol.* **2017**, *8*, 1–18. [CrossRef]

63. Paulo, B.S.; Sigrist, R.; Angolini, C.F.F.; Oliveira, L.G. De New cyclodepsipeptide derivatives revealed by genome mining and molecular networking. *Chem. Sel.* **2019**, *4*, 7785–7790.

64. Purves, K.; Macintyre, L.; Brennan, D.; Hreggviðsson, G.Ó.; Kuttner, E.; Ásgeirsdóttir, M.E.; Young, L.C.; Green, D.H.; Edrada-ebel, R.; Duncan, K.R. Using molecular networking for microbial secondary metabolite bioprospecting. *Metabolites* **2016**, *6*, 2. [CrossRef]

65. Deng, H.; Ma, L.; Bandaranayaka, N.; Qin, Z.; Mann, G.; Kyeremeh, K.; Yu, Y.; Shepherd, T.; Naismith, J.H.; O'Hagan, D. Identification of fluorinases from *Streptomyces* sp. MA37, *Norcardia brasiliensis*, and *Actinoplanes* sp. N902-109 by genome mining. *ChemBioChem* **2014**, *15*, 364–368. [CrossRef] [PubMed]

66. Anoop, A.; Antunes, A. Whole genome sequencing of the symbiont *Pseudovibrio* sp. from the intertidal marine sponge *Polymastia penicillus* revealed a gene repertoire for host-switching permissive lifestyle. *Genome Biol. Evol.* **2015**, *7*, 3022–3032.

67. Bertin, M.J.; Schwartz, S.L.; Lee, J.; Korobeynikov, A.; Dorrestein, P.C.; Gerwick, L.; Gerwick, W.H. Spongosine production by a *Vibrio harveyi* strain associated with the sponge *Tectitethya crypta*. *J. Nat. Prod.* **2014**, *78*, 493–499. [CrossRef]

68. Jeske, O.; Jogler, M.; Petersen, J.; Sikorski, J.; Jogler, C. From genome mining to phenotypic microarrays: *Planctomycetes* as source for novel bioactive molecules. *Antonie Van Leeuwenhoek* **2013**, *104*, 551–567. [CrossRef]

69. Guérard-Hélaine, C.; de Berardinis, V.; Besnard-Gonnet, M.; Darii, E.; Debacker, M.; Debard, A.; Fernandes, C.; Hélaine, V.; Mariage, A.; Pellouin, V.; et al. Genome mining for innovative biocatalysts: New dihydroxyacetone aldolases for the Chemist's Toolbox. *ChemCatChem* **2015**, *7*, 1871–1879. [CrossRef]

70. Singh, S.P.; Klisch, M.; Sinha, R.P.; Häder, D. Genomics genome mining of mycosporine-like amino acid (MAA) synthesizing and non-synthesizing cyanobacteria: A bioinformatics study. *Genomics* **2010**, *95*, 120–128. [CrossRef]

71. Micallef, M.L.; Agostino, P.M.D.; Sharma, D.; Viswanathan, R.; Moffitt, M.C. Genome mining for natural product biosynthetic gene clusters in the Subsection V cyanobacteria. *BMC Genom.* **2015**, 1–20. [CrossRef]

72. Fuerst, J.A.; Sagulenko, E. Beyond the bacterium: Planctomycetes structure and function. *Nat. Rev. Microbiol.* **2011**, *9*, 13–18. [CrossRef]

73. Oren, A.; Gunde-cimerman, N. Mycosporines and mycosporine-like amino acids: UV protectants or multipurpose secondary metabolites? *FEMS Microbiol. Lett.* **2007**, *269*, 1–10. [CrossRef]

74. Keller, N.P.; Turner, G.; Bennett, J.W. Fungal secondary metabolism—from biochemistry to genomics. *Nat. Rev. Microbiol.* **2015**, *3*, 937–947. [CrossRef]

75. Bergmann, S.; Schümann, J.; Scherlach, K.; Lange, C.; Brakhage, A.A.; Hertweck, C. Genomics-driven discovery of PKS-NRPS hybrid metabolites from *Aspergillus nidulans*. *Nat. Chem. Biol.* **2007**, *3*, 213–217. [CrossRef]

76. Mao, X.; Xu, W.; Li, D.; Yin, W.; Chooi, Y.; Li, Y.; Tang, Y.; Hu, Y. Epigenetic genome mining of an endophytic fungus leads to the pleiotropic biosynthesis of natural products. *Angew. Commun.* **2015**, *54*, 7592–7596. [CrossRef]

77. Ding, Y.; De Wet, J.R.; Cavalcoli, J.; Li, S.; Greshock, T.J.; Miller, K.A.; Finefield, J.M.; Sunderhaus, J.D.; Mcafoos, T.J.; Tsukamoto, S.; et al. Genome-based characterization of two prenylation steps in the assembly of the Stephacidin and Notoamide anticancer agents in a marine-derived *Aspergillus* sp. *J. Am. Chem. Soc.* **2010**, *132*, 12733–12740. [CrossRef]

78. Ye, Y.; Minami, A.; Mándi, A.; Liu, C.; Taniguchi, T.; Kuzuyama, T.; Monde, K.; Gomi, K.; Oikawa, H. Genome mining for sesterterpenes using bifunctional terpene synthases reveals a unified intermediate of di/sesterterpenes. *J. Am. Chem. Soc.* **2015**, *137*, 11846–11853. [CrossRef]

79. Mojib, N.; Amad, M.; Thimma, M.; Aldanondo, N.; Kumaran, M.; Irigoien, X. Carotenoid metabolic profiling and transcriptome-genome mining reveal functional equivalence among blue-pigmented copepods and appendicularia. *Mol. Ecol.* **2014**, *23*, 2740–2756. [CrossRef]

80. Huang, A.C.; Kautsar, S.A.; Hong, Y.J.; Medema, M.H.; Bond, A.D.; Tantillo, D.J.; Osbourn, A. Unearthing a sesterterpene biosynthetic repertoire in the Brassicaceae through genome mining reveals convergent evolution. *Proc. Natl. Acad. Sci. USA* **2017**, *114*, 1–10. [CrossRef]

 *marine drugs*

Article

Cloning, Expression, and Characterization of a New PL25 Family Ulvan Lyase from Marine Bacterium *Alteromonas* sp. A321

Jian Gao, Chunying Du, Yongzhou Chi, Siqi Zuo, Han Ye and Peng Wang *

College of Food Science and Engineering, Ocean University of China, Qingdao 266003, China
* Correspondence: pengwang@ouc.edu.cn; Tel.: +86-0532-82032290

Received: 25 August 2019; Accepted: 1 October 2019; Published: 8 October 2019

Abstract: Ulvan lyases can degrade ulvan to oligosaccharides with potent biological activity. A new ulvan lyase gene, *ALT3695*, was identified in *Alteromonas* sp. A321. Soluble expression of *ALT3695* was achieved in *Escherichia coli* BL21 (DE3). The 1314-bp gene encoded a protein with 437 amino acid residues. The amino acid sequence of ALT3695 exhibited low sequence identity with polysaccharide lyase family 25 (PL25) ulvan lyases from *Pseudoalteromonas* sp. PLSV (64.14% identity), *Alteromonas* sp. LOR (62.68% identity), and *Nonlabens ulvanivorans* PLR (57.37% identity). Recombinant ALT3695 was purified and the apparent molecular weight was about 53 kDa, which is different from that of other polysaccharide-degrading enzymes identified in *Alteromonas* sp. A321. ALT3695 exhibited maximal activity in 50 mM Tris-HCl buffer at pH 8.0 and 50 °C. ALT3695 was relatively thermostable, as 90% activity was observed after incubation at 40 °C for 3 h. The K_m and V_{max} values of ALT3695 towards ulvan were 0.43 mg·mL^{-1} and 0.11 μmol·min^{-1}·mL^{-1}, respectively. ESI-MS analysis showed that *enzymatic products* were mainly disaccharides and tetrasaccharides. This study reports a new PL25 family ulvan lyase, ALT3695, with properties that suggest its great potential for the preparation of ulvan oligosaccharides.

Keywords: ulvan-derived oligosaccharides; ulvan lyase; heterologous expression; polysaccharide lyase family 25

1. Introduction

Ulvan is a type of cell wall polysaccharide, extracted from marine green seaweed (genus *Ulva* and *Enteromorpha*) [1]. It is mainly composed of iduronic acid (IdoA), 3-sulfated rhamnose (Rha3S), xylose (Xyl), and glucuronic acid (GlcA) [1]. The repetitive disaccharide units in ulvan are Rha3S-GlcA, Rha3S-IdoA, and Rha3S-Xyl [1]. Ulvan has been shown to possess various pharmacological properties, such as antioxidant [2], anticoagulant [3], antitumor [4], antihyperlipidemic [5], and antiviral [6] activities. However, its biological activity is greatly affected by its molecular weight, and its application could be limited by its high viscosity [3,7,8]. Compared to high-molecular-weight ulvan, low-molecular-weight ulvan possesses higher solubility, lower viscosity, easier absorption, and greater exposure of reactive groups [3,5,9]. In addition, it has been reported that low-molecular-weight sulfated polysaccharides [7] and their iron complexes [10] have potent biological activity. Thus, ulvan-derived oligosaccharides have been attracting increased attention for their tremendous potential for applications in the functional food and pharmaceutical industries.

Several methods for depolymerizing polysaccharides have been studied. Qi et al. prepared different molecular weight ulvans by treatment with H_2O_2 and by changing the depolymerization conditions [7]. Yu et al. degraded ulvan by microwaving under different pressures [8]. Mild acid hydrolysis has also been used to degrade ulvan [1]. However, such chemical and physical methods have many shortcomings, including sulfate group loss, complex products, and low oligosaccharide yields [9]. In contrast, enzymatic depolymerization is a potentially energy-efficient and environmentally friendly method, with no requirement for harsh reagents or extreme conditions [11]. Moreover, specific enzymes can be very useful for studying polysaccharide structure [12,13]. Therefore, it is necessary to find new enzymes that are capable of degrading ulvan, such as ulvan lyases.

The first marine bacterium capable of degrading ulvan was isolated by Lahaye et al. [12]. However, no in-depth study of its ulvan-degrading enzymes has been reported. Recently, Collén et al. reported the first endolytic ulvan lyase genes [14], which were found in the ulvanolytic bacterium *Nonlabens ulvanivorans*. Subsequently, the genomes of three ulvanolytic bacteria were sequenced [15], and ulvan lyases belonging to different polysaccharide lyase families were found. Kopel et al. reported several ulvan lyases in these three genomes [15], and these ulvan lyases showed no homology to those found by Collén et al. [14], which belong to PL24 family. Foran et al. identified another novel ulvan lyase (LOR_29) in the *Alteromonas* sp. LOR genome, which is the founding member of polysaccharide lyase family 25 (PL25) [16]. Thus far, three ulvan lyase families have been established (http://www.cazy.org), including PL24, PL25, and PL28. Structural characterizations of representative enzymes from these three families have also been reported [17–19]. As the primary ulvan-degrading enzyme [16], ulvan lyase catalyzes β-elimination at the internal bond between uronic acid and Rha3S, producing oligosaccharides with unsaturated uronic acid (ΔGlcA) [14,15]. Compared to other methods, the uniform enzymatic product is an advantage of using ulvan lyases to degrade ulvan, which is convenient for studying their pharmacological activity. In addition, sulfate groups are well retained during the degradation process, which is essential for the activity of ulvan oligosaccharides [9]. Ulvan lyases have also been used for epitope deletion studies [20]. However, only seven ulvan lyases have been characterized. To expand the repertoire of enzymes to efficiently produce ulvan-derived oligosaccharides, additional new ulvan lyases must be investigated.

Previous studies showed that *Alteromonas* sp. A321 was capable of degrading ulvan [9]. In this study, a new ulvan lyase gene, *ALT3695*, was identified in *Alteromonas* sp. A321 and soluble expression of ALT3695 was achieved in *Escherichia coli* BL21 (DE3). Recombinant ulvan lyases were purified and the molecular weight was investigated. ALT3695 differs from other enzymes previously found in *Alteromonas* sp. A321 [21]. Thus, this study reports a new enzyme for preparing ulvan-derived oligosaccharides and enriches the marine enzyme library.

2. Results and Discussion

2.1. Sequence Analysis

The *ALT3695* gene is 1314 bp in length and encodes a 437-amino acid protein. The ALT3695 amino acid sequence shares 64.14%, 62.68%, and 57.37% sequence identity with reported ulvan lyases from *Pseudoalteromonas* sp. PLSV (PLSV_3936, GenBank accession no. WP_033186995.1) [18], *Alteromonas* sp. LOR (LOR_29, GenBank accession no. WP_052010178.1) [16], and *Nonlabens ulvanivorans* PLR (NLR_492, GenBank accession no. WP_036580476.1) [16], respectively. As the representative enzyme of PL25 family, the structure and catalytic mechanism of PLSV_3936 have been investigated [18]. In PLSV_3936, His123, His143, Tyr 188, Arg204, and Tyr246 are conserved and have been proposed as active site residues, and Gln66, Tyr246, and Arg282 are highly conserved.

Several homologous enzymes from different organisms with less sequence identity were selected in the carbohydrate-active enzymes (CAZy) database. Amino acid sequence alignment showed that most residues are also conserved in ALT3695 and other PL25 family members, except Gln66 (Figure 1). Among these residues, His143 and Try246 could help Arg204 to neutralize the negative charge on glucuronic acid. His123 and Tyr188 were acid-base catalysis residues [18]. A phylogenetic tree of ALT3695 and other reported ulvan lyases was constructed by the neighbor-joining method, which suggested that ALT3695 is a PL25 family ulvan lyase (Figure 2).

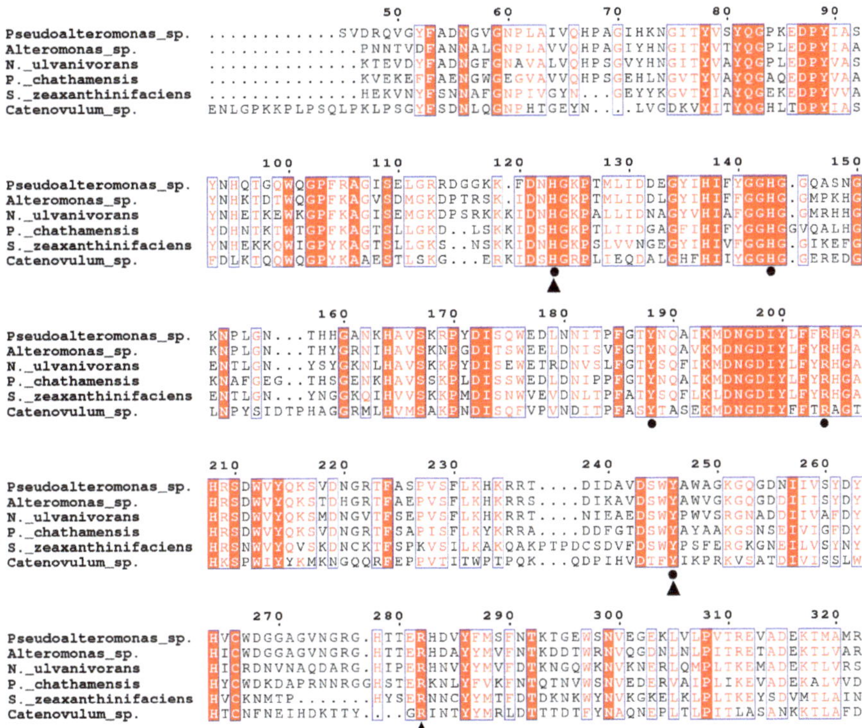

Figure 1. Amino acid sequence alignment of ALT3695 with ulvan lyases from *Pseudoalteromonas* sp. PLSV (PLSV_3936, GenBank accession no. WP_033186995.1), *Alteromonas* sp. A321 (ALT3695, GenBank accession no. MN347032), *Nonlabens ulvanivorans* PLR (NLR_492, GenBank accession no. WP_036580476.1), *Paraglaciecola chathamensis* S18K6 (GCHA_4617, GenBank accession no. GAC12534.1), *Siansivirga zeaxanthinifaciens* CC-SAMT-1 (AW14_13480, GenBank accession no. AJR04515.1), and *Catenovulum* sp. CCB-QB4 (C2869_03520, GenBank accession no. AWB65560.1). Active site residues are marked with filled circles (●). Highly conserved residues are marked with filled triangles (▲).

2.2. Expression and Purification of Recombinant ALT3695

Soluble expression of His-tagged ALT3695 ulvan lyase was achieved in *E. coli* BL21 (DE3) by adding 1 mM isopropyl-β-d-thiogalactopyranoside (IPTG). The recombinant ALT3695 was purified, and SDS-PAGE showed only a single protein band. The purified ALT3695 showed a specific activity of 1.88 U/mg protein, with 52% recovery (Table 1). As expected, the apparent molecular weight was 53 kDa (Figure 3), which is similar to that previously reported for PL25 family ulvan lyases, such as LOR_29 (52 kDa) [16] and NLR_492 (55 kDa) [16]. The molecular weight of ulvan lyases from other families, such as IL45_01510 (GenBank accession no. AEN28574.1, PL28) [14] and PsPL (GenBank accession no. AMA19992.1, PL24) [22], were about 46 kDa and 59 kDa, respectively.

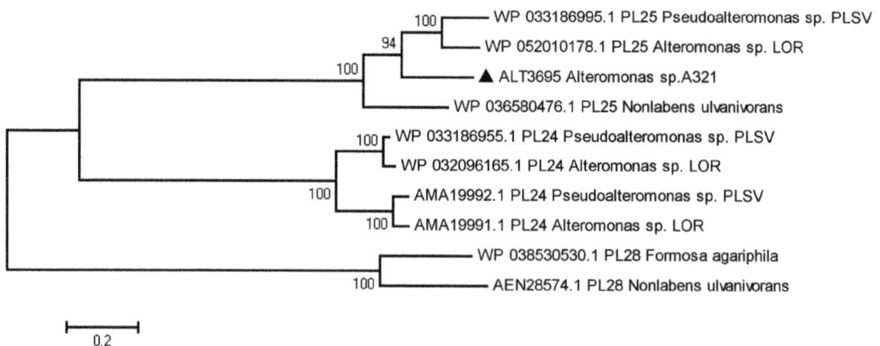

Figure 2. Phylogenetic tree of ALT3695 (filled triangle) and other ulvan lyases generated using the neighbor-joining method. Numbers along the branch nodes represent bootstrap percentages based on 1000 resamplings. The scale bar indicates the average number (0.2) of amino acid substitutions per site. *Pseudoalteromonas* sp. PLSV (PLSV_3936, GenBank accession no. WP_033186995.1), *Alteromonas* sp. LOR (LOR_29, GenBank accession no. WP_052010178.1), *Nonlabens ulvanivorans* (NLR_492, GenBank accession no. WP_036580476.1), *Pseudoalteromonas* sp. PLSV (PLSV_3925, GenBank accession no. WP_033186955.1), *Alteromonas* sp. LOR (LOR_107, GenBank accession no. AMA19991.1), *Pseudoalteromonas* sp. PLSV (PLSV_3875, GenBank accession no. AMA19992.1), *Alteromonas* sp. LOR (LOR_61, GenBank accession no. WP_032096165.1), *Formosa agariphila* KMM 3901 (BN863_22190, GenBank accession no. WP_038530530.1), and *Nonlabens ulvanivorans* PLR (IL45_01510, GenBank accession no. AEN28574.1).

Table 1. Characterization of purified ALT3695.

Purification Steps	Total Protein (mg)	Total Activity (U)	Specific Activity (U/mg)	Purification (fold)	Yield (%)
Crude enzymes	21.39	20.49	0.96	1.00	100.00
BeaverBeads™ IDA-Nickel	5.69	10.68	1.88	1.96	52.12

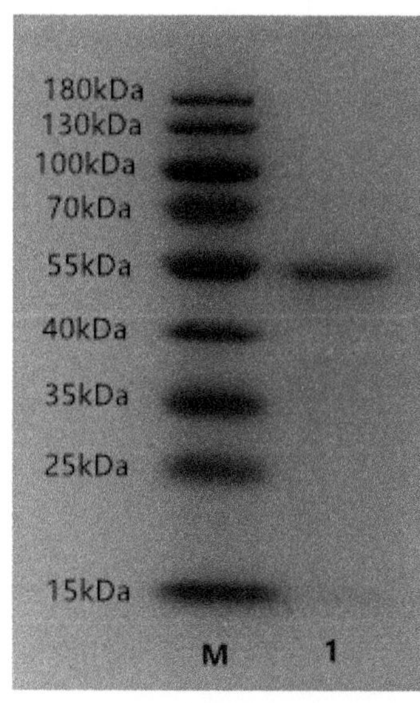

Figure 3. SDS-PAGE of purified ALT3695. Lane 1: purified ALT3695; Lane M: molecular weight marker.

2.3. Influence of Temperature on the Activity of Recombinant ALT3695

The influence of temperature on enzyme stability and the optimum temperature for ALT3695 activity were investigated. Ulvan lyases from different families exhibit different optimum temperatures. The optimum temperature for ALT3695 activity was 50 °C (Figure 4a), which is higher than that of another ulvan lyase from the PL25 family (45 °C) [16]. Ulvan lyases from other families, such as FaPL28 (GenBank accession no. WP_038530530.1, PL28) [23] and PsPL (GenBank accession no. AMA19992.1, PL24) [22], showed maximum activity at lower temperatures of 29.5 °C and 35 °C, respectively. Little has been reported concerning the thermal stability of PL25 family members. FaPL28 had poor thermal stability, and less than 10% activity was observed after incubation for 3 h at or above 30.9 °C [23]. In contrast, ALT3695 was relatively thermostable and 90% activity was observed even after incubation for 3 h at 40 °C (Figure 4b). However, the stability of ALT3695 decreased as the temperature was increased above 40 °C (Figure 4b), 49% residual activity was observed after incubation for 3 h at 45 °C, and no activity was observed after incubation at 55 °C for 1 h.

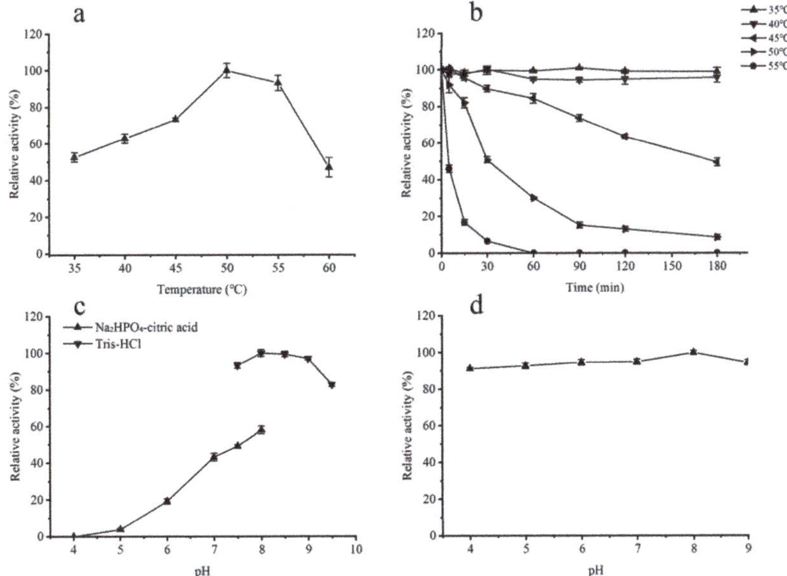

Figure 4. Biochemical characterization of ALT3695. (**a**) Activity of ALT3695 over a range of temperatures (35–60 °C) to determine the optimal temperature. (**b**) Thermal stability of ALT3695. (**c**) Activity of ALT3695 over a range of pH values to determine the optimal pH. (**d**) The pH stability of ALT3695.

2.4. Influence of pH on the Activity of Recombinant ALT3695

To investigate the effect of pH on ALT3695 activity, different buffers (pH 4–9.5) were used. The optimum pH of other ulvan lyases ranged from 7.5 to 9.0 [16,22,23], which corresponds to the pH of slightly alkaline seawater [24]. Similarly, ALT3695 showed maximal activity at pH 8.0 (Figure 4c). In addition, activity in Tris-HCl buffer was higher than that in Na_2HPO_4-citric acid buffer at pH 8.0, suggesting that the catalytic activity of ALT3695 was also affected by the buffer salt. ALT3695 retained 80% activity after pre-incubation at pH 4.0–9.0 for 2 h at 4 °C, indicating that ALT3695 has excellent pH stability (Figure 4d).

2.5. Influences of Surfactants, Metal Ions, and Metal Chelators on the Activity of Recombinant ALT3695

Table 2a shows the influence of various metal ions on ALT3695 activity. K^+ had no significant influence on ALT3695 activity. Additionally, Fe^{2+} and Cu^{2+} decreased enzyme activity, and this effect was also found in the ulvan lyases, PsPL [22] and FaPL28 [23]. Furthermore, ulvan lyases from different families exhibit different responses to Co^{2+}, the activity of PsPL was increased by Co^{2+} [22], while the activity of FaPL28 [23] and ALT3695 was decreased by Co^{2+}. This may be attributed to structural differences among these enzyme families. Moreover, Cd^{2+}, Hg^{2+}, and Fe^{3+} deactivated ALT3695 completely, which may be caused by the binding of these ions to the SH, CO, and NH moieties of amino acids, resulting in structural changes and inactivation [21]. It is worth mentioning that although ALT3695 has Zn^{2+}-containing ligands, Zn^{2+} strongly inhibited enzymatic activity. Thus, Zn^{2+} may play a dual role, as inherent Zn^{2+} is essential for maintaining enzyme structure, but exogenous Zn^{2+} may bind to other amino acid residues and inhibit activity [25].

Table 2. (a) Effects of various metal ions on the activity of ALT3695 ulvan lyase. (b) Effects of surfactants and metal chelators on enzyme activity.

		(a)		
Metal Ions	**Concentration (mM)**	**Relative Activity (%)**	**Concentration (mM)**	**Relative Activity (%)**
Control	-	100.01 ± 1.37	-	100.01 ± 1.37
K^+	10	101.85 ± 1.66	20	103.62 ± 1.92
Ca^{2+}	10	112.34 ± 0.48	20	142.51 ± 2.62
Mg^{2+}	10	106.13 ± 0.42	20	117.13 ± 2.12
Zn^{2+}	10	ND	20	ND
Ba^{2+}	10	105.49 ± 2.03	20	121.23 ± 2.39
Cu^{2+}	10	17.05 ± 0.68	20	ND
Fe^{2+}	10	19.65 ± 2.85	20	ND
Co^{2+}	10	29.88 ± 2.51	20	ND
Cd^{2+}	10	ND	20	ND
Hg^{2+}	10	ND	20	ND
Fe^{3+}	10	ND	20	ND
		(b)		
Reagents	**Concentration (mM)**	**Relative Activity (%)**	**Concentration (mM)**	**Relative Activity (%)**
Control	-	100 ± 0.21	-	100 ± 0.21
Tween-20	5	98.73 ± 2.22	10	84.17 ± 1.52
Tween-80	5	103.29 ± 0.44	10	101.14 ± 1.67
Triton X-100	5	104.01 ± 0.32	10	97.36 ± 1.85
EDTA	5	32.11 ± 1.23	10	ND
1,10-phenanthroline	5	106.71 ± 0.79	10	117.08 ± 1.71

ND: No activity was determined.

The enzyme activity was increased by Ca^{2+}, Mg^{2+}, and Ba^{2+}. The addition of Ca^{2+} would not only increase the activity of PL25 ulvan lyase, but also stimulate the activity of PsPL [22] and FaPL28 [23]. With more Ca^{2+} added, the activity of ALT3695 was also increased. In addition, ALT3695 activity decreased by 68% after pre-incubation in 5 mM ethylenediaminetetraacetic acid (EDTA), similar phenomenon was also found in PsPL and FaPL28, this may due to the removal of divalent cations bound to enzymes. Additionally, structure analysis suggested that divalent metal ions play a structural role in all three ulvan lyase families.

The effect of surfactants and metal chelators on ALT3695 activity was also investigated (Table 2b). It has been reported that non-ionic surfactants could increase the activity of some enzymes [26,27]. However, surfactants had little effect on the activity of ALT3695.

2.6. Kinetic Parameters of Recombinant ALT3695

V_{max} is the maximum initial rate of an enzymatic reaction. Each enzyme has a specific K_m for each substrate, which is inversely related to the enzyme's affinity for the substrate [28]. The kinetic parameters of other ulvan lyases from the PL25 family have not yet been reported. The K_m and V_{max} of recombinant ALT3695 toward ulvan were 0.43 mg·mL^{-1} and 0.11 μmol·min^{-1}·mL^{-1}, respectively, as determined by the Lineweaver-Burk plot. ALT3695 showed a higher affinity for ulvan than PsPL and FaPL28, which had a K_m of 2.10 mg·mL^{-1} [22] and 0.75 mg·mL^{-1} [23], respectively, suggesting the potential of ALT3695 for application.

2.7. Analysis of Enzymatic Products

The main repetitive disaccharide units in ulvan are Rha3S-GlcA, Rha3S-IduA, and Rha3S-Xyl. ALT3695 could cleave the bond on both Rha3S-GlcA and Rha3S-IdoA, while PL24 family ulvan lyase specifically cleaves the bond between Rha3S-GlcA. The products of ulvan degradation by ALT3695 were analyzed by negative electrospray ionisation mass spectrometry (ESI-MS). Mass spectroscopic analysis of products showed two major types of oligosaccharides. The peak at m/z 401 was equivalent to the disaccharide ΔGlcA-Rha3S, with a molecular weight of 402 Da. There was an unsaturated double bond between C4 and C5 [18]. The peak at m/z 379 was equivalent to double-charged molecular ions of the tetrasaccharide ΔGlcA-Rha3S-Xyl-Rha3S, with a molecular weight of 760 Da [14]. A low abundance of single-charged molecular ions of this tetrasaccharide was also identified as the peak at m/z 759. However, the products of ulvan degradation by PL24 ulvan lyase were different. The products included the two most abundant oligosaccharides which comprised of disaccharides (ΔGlcA-Rha3S) and tetrasaccharides (ΔGlcA-Rha3S-IdoA-Rha3S), with minor tetrasaccharides (ΔGlcA-Rha3S-Xyl-Rha3S) [15].

Moreover, the change in the full wavelength during enzymatic degradation was studied (Figure 5b). As the enzymatic reaction continued, the absorption peak at 235 nm gradually increased, which may be due to the generation of ΔGlcA-Rha3S and ΔGlcA-Rha3S-Xyl-Rha3S, as C-4=C-5 double bonds have a high absorbance at 235 nm [29].

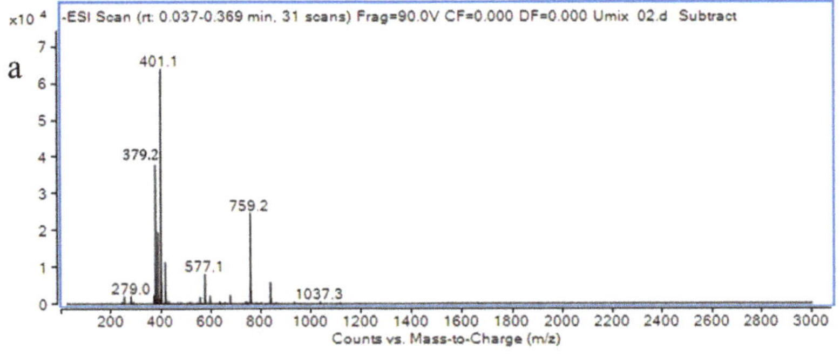

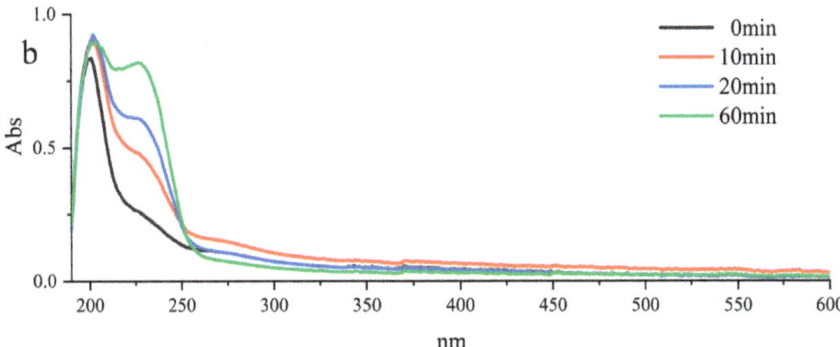

Figure 5. Analysis of the enzymatic products of ulvan generated by ALT3695. (**a**) ESI-MS analysis of products. (**b**) Full wavelength scans conducted during enzymatic degradation.

3. Materials and Methods

3.1. Strains, Plasmids, and Medium

The ulvan-degrading strain *Alteromonas* sp. A321 was provided by Peng Wang (Ocean University of China, Qingdao, China). Genomic DNA was extracted, and the whole genome was sequenced. *E. coli* strains DH5α and BL21 (DE3) were purchased from TIANGEN Biotech Co. (Beijing, China). The pProEX-HTa vector was used to clone and express the ulvan lyase gene. The *E. coli* strains were cultured in Luria-Bertani medium.

3.2. Sequence Analysis

An ulvan lyase gene, named *ALT3695*, was identified in the genome, and its sequence was deposited into the NCBI database (GenBank accession no. MN347032). DNAMAN 9 (Lynnon Biosoft, San Ramon, CA, USA) software was used to analyze the sequence of *ALT3695*. The homology was analyzed using BLAST at the NCBI website. A phylogenetic tree based on the homology of various ulvan lyases was constructed using MEGA 5.0 (Koichiro, Tokyo, Japan).

3.3. Construction of the Recombinant ALT3695-Expressing Strain

The *ALT3695* gene was amplified by PCR using Phanta® Turbo Super-Fidelity DNA Polymerase (Vazyme Biotech, Nanjing, China). Two primers were designed to amplify the *ALT3695* gene, 5′-CCGGAATTCCGGCCCAATAATACTGTCGACTTCGCTAACA-3′ and 5′-GGGGTACCCCTTACTTATTCTCGCTTCGTATTGGTCC-3′. Then, the amplicon was purified, digested, and ligated into the pProEX-HTa vector to generate pProEX-HTa-ALT3695. The recombinant plasmid pProEX-HTa-ALT3695 was transformed into *E. coli* DH5α for amplification, and then transformed into *E. coli* BL21(DE3) for expression. Positive strains were verified by sequencing.

3.4. Expression of Recombinant ALT3695

E. coli BL21 (DE3) cells containing pProEX-HTa-ALT3695 were cultivated in Luria-Bertani medium supplemented with ampicillin (100 µg/mL). The culture conditions were 37 °C and 200 rpm. When the strain entered logarithmic growth phase (OD600 ~ 0.7), IPTG was added at a final concentration of 1 mM to induce the expression of ALT3695. Then, the culture temperature was decreased to 28 °C, but 200 rpm was retained with further cultivation for 18 h. Finally, cells were collected by centrifugation under freezing conditions (4 °C) and resuspended in 20 mM Tris-HCl buffer (pH 7.5).

3.5. Enzyme Purification

After homogenization by sonication, the supernatant containing crude enzymes was obtained by centrifugation under freezing conditions. The His-tagged ulvan lyase was purified using BeaverBeads™ IDA-Nickel (Beaverbio, Suzhou, China) and eluted using 20 mM Tris-HCl buffer (pH 7.5) supplemented with 0.5 M NaCl and 300 mM imidazole. The concentration of soluble protein was determined by the Folin-Ciocalteu method [30]. Purified enzyme was then subjected to SDS-PAGE to analyze the purity and molecular mass [31].

3.6. Enzyme Activity Assay

Ulvan was extracted from *Ulva clathrate* (Fujian, China) according to the method of Qi et al. [32]. The ulvan substrate was added at 0.1% (w/v) in 50 mM Tris-HCl buffer (pH 8.0) and the reaction was conducted at 35 °C. The activity of ulvan lyase was examined by measuring the change in absorbance at 235 nm [14]. One unit (U) of ulvan lyase activity was defined as the amount of protein needed to generate 1 µmol of unsaturated glucuronyl residue (extinction coefficient, 4800 $M^{-1} \cdot cm^{-1}$) per minute [23]. All assays were repeated three times.

3.7. Characterization of Recombinant ALT3695

To examine the optimum temperature of recombinant ALT3695, the reaction was conducted at 35–60 °C under standard conditions. The change in enzyme stability at different temperatures was analyzed by conducting enzyme activity assays after pre-incubating the enzyme for various times (5, 15, 30, 60, 90, 120, and 180 min) at 35–60 °C.

The optimal pH of recombinant ALT3695 was investigated by examining its activity in 50 mM Na_2HPO_4-citric acid buffer (for pH 4.0–8.0) and 50 mM Tris-HCl buffer (for pH 7.5–9.5). To investigate the change of enzyme stability by pH, purified enzymes were stored for 2 h at 4 °C in buffers of different pH levels, and then assayed for residual enzyme activity.

To examine the influence of different metal ions on recombinant ALT3695 activity, the enzyme was incubated with various metal ions (at 10 mM and 20 mM) for 2 h at 4 °C. The following metal ions were tested: K^+, Ca^{2+}, Mg^{2+}, Zn^{2+}, Ba^{2+}, Cu^{2+}, Fe^{2+}, Hg^{2+}, Co^{2+}, Cd^{2+}, and Fe^{3+}. The effect of the following surfactants and metal chelators on recombinant ALT3695 activity was also determined: Tween-20, Tween-80, Triton X-100, EDTA, and 1,10-phenanthroline. The enzyme solution was stored with each surfactant or potential inhibitor (at 5 mM and 10 mM) for 2 h at 4 °C. Then, enzyme activity was assayed and compared with the activity in the absence of additive.

3.8. Kinetic Measurements

To determine the kinetic parameters of recombinant ALT3695 using ulvan as a substrate, reactions were performed with various concentrations (0.03125–1.0 mg·mL^{-1}) of substrate under standard conditions for 5 min. Then, the K_m and V_{max} values were calculated by constructing a Lineweaver-Burk double reciprocal plot.

3.9. Analysis of Enzymatic Products

To obtain enzymatic products for analysis, purified ALT3695 was incubated with ulvan (5 mg·mL^{-1}) for 3 h at 35 °C. Then, product samples were collected, and the product molecular weights were measured by negative-ion ESI-MS (Agilent 1290 Infinity II-6460, Agilent Corp., Wilmington, DE, USA; Frag = 90.0 V, m/z 100–3000 amu). Full wavelength scans (190–600 nm) were obtained during enzymatic degradation.

4. Conclusions

ALT3695 is a new ulvan lyase identified from *Alteromonas* sp. A321. The gene was cloned, and the recombinant ALT3695 was expressed in soluble fraction. The enzymatic properties were also investigated. The enzyme exhibited excellent stability and substrate affinity. Maximum activity was observed at 50 °C, and 90% activity remained after incubation at 40 °C for 3 h. ALT3695 catalyzes β-elimination at the internal bonds between uronic acids and Rha3S, leading to increased absorbance at 235 nm. ESI-MS results showed that disaccharides and tetrasaccharides were the major enzymatic products. In conclusion, ALT3695 shows great potential for the preparation of ulvan oligosaccharides. Further research on this recombinant strain and a structural analysis of ulvan lyase ALT3695 are warranted based on results of this study.

Author Contributions: J.G., C.D., and P.W. conceived and designed the experiments; J.G., C.D., and Y.C. performed the experiments; S.Z. and H.Y. analyzed the data; J.G. and P.W. wrote the paper.

Funding: This research was funded by the Key Research and Development Program of Shandong 2018GHY115032, the National Key Research and Development Program 2018YFC0311203 and the Innovation and Development of Marine Economy Demonstration City (Weihai) Program.

Conflicts of Interest: The authors declare no conflict of interest.

References

1. Lahaye, M.; Robic, A. Structure and functional properties of ulvan, a polysaccharide from green seaweeds. *Biomacromolecules* **2007**, *8*, 1765–1774. [CrossRef] [PubMed]

2. Li, W.; Wang, K.; Jiang, N.; Liu, X.; Wan, M.; Chang, X.; Liu, D.; Qi, H.; Liu, S. Antioxidant and antihyperlipidemic activities of purified polysaccharides from *Ulva pertusa*. *J. Appl. Phycol.* **2018**, *30*, 2619–2627. [CrossRef]

3. Cui, J.; Li, Y.; Wang, S.; Chi, Y.; Hwang, H.; Wang, P. Directional preparation of anticoagulant-active sulfated polysaccharides from *Enteromorpha prolifera* using artificial neural networks. *Sci. Rep.* **2018**, *8*, 3062. [CrossRef]

4. Thanh, T.T.; Quach, T.M.; Nguyen, T.N.; Vu Luong, D.; Bui, M.L.; Tran, T.T. Structure and cytotoxic activity of ulvan extracted from green seaweed *Ulva lactuca*. *Int. J. Biol. Macromol.* **2016**, *93*, 695–702. [CrossRef] [PubMed]

5. Qi, H.; Huang, L.; Liu, X.; Liu, D.; Zhang, Q.; Liu, S. Antihyperlipidemic activity of high sulfate content derivative of polysaccharide extracted from *Ulva pertusa* (Chlorophyta). *Carbohydr. Polym.* **2012**, *87*, 1637–1640. [CrossRef]

6. Aguilar-Briseño, J.A.; Cruz-Suarez, L.E.; Sassi, J.F.; Ricque-Marie, D.; Zapata-Benavides, P.; Mendoza-Gamboa, E.; Rodríguez-Padilla, C.; Trejo-Avila, L.M. Sulphated polysaccharides from *Ulva clathrata* and *Cladosiphon okamuranus* seaweeds both inhibit viral attachment/entry and cell-cell fusion, in NDV infection. *Mar. Drugs* **2015**, *13*, 697–712. [CrossRef]

7. Qi, H.; Zhao, T.; Zhang, Q.; Li, Z.; Zhao, Z.; Xing, R. Antioxidant activity of different molecular weight sulfated polysaccharides from *Ulva pertusa* Kjellm (Chlorophyta). *J. Appl. Phycol.* **2005**, *17*, 527–534. [CrossRef]

8. Yu, P.; Li, N.; Liu, X.; Zhou, G.; Zhang, Q.; Li, P. Antihyperlipidemic effects of different molecular weight sulfated polysaccharides from *Ulva pertusa* (Chlorophyta). *Pharmacol. Res.* **2003**, *48*, 543–549.

9. Li, Y.; Wang, J.; Yu, Y.; Li, X.; Jiang, X.; Hwang, H.; Wang, P. Production of enzymes by *Alteromonas* sp. A321 to degrade polysaccharides from *Enteromorpha prolifera*. *Carbohydr. Polym.* **2013**, *98*, 988–994. [CrossRef]

10. Cui, J.; Li, Y.; Yu, P.; Zhan, Q.; Wang, J.; Chi, Y.; Wang, P. A novel low molecular weight *Enteromorpha* polysaccharide-iron (III) complex and its effect on rats with iron deficiency anemia (IDA). *Int. J. Biol. Macromol.* **2018**, *108*, 412–418. [CrossRef]

11. Li, Y.; Huang, Z.; Qiao, L.; Gao, Y.; Guan, H.; Hwang, H.; Aker, W.G.; Wang, P. Purification and characterization of a novel enzyme produced by *Catenovulum* sp. LP and its application in the pre-treatment to *Ulva prolifera* for bio-ethanol production. *Process Biochem.* **2015**, *50*, 799–806. [CrossRef]

12. Lahaye, M.; Brunel, M.; Bonnin, E. Fine chemical structure analysis of oligosaccharides produced by an ulvan-lyase degradation of the water-soluble cell-wall polysaccharides from *Ulva* sp. (Ulvales, Chlorophyta). *Carbohydr. Res.* **1997**, *304*, 325–333. [CrossRef]

13. Konasani, V.R.; Jin, C.; Karlsson, N.G.; Albers, E. Ulvan lyase from *Formosa agariphila* and its applicability in depolymerisation of ulvan extracted from three different *Ulva* species. *Algal Res.* **2018**, *36*, 106–114. [CrossRef]

14. Collén, P.N.; Sassi, J.F.; Rogniaux, H.; Marfaing, H.; Helbert, W. Ulvan lyases isolated from the Flavobacteria *Persicivirga ulvanivorans* are the first members of a new polysaccharide lyase family. *J. Biol. Chem.* **2011**, *286*, 42063–42071. [CrossRef] [PubMed]

15. Kopel, M.; Helbert, W.; Belnik, Y.; Buravenkov, V.; Herman, A.; Banin, E. New family of ulvan lyases identified in three isolates from the alteromonadales order. *J. Biol. Chem.* **2016**, *291*, 5871–5878. [CrossRef] [PubMed]

16. Foran, E.; Buravenkov, V.; Kopel, M.; Mizrahi, N.; Shoshani, S.; Helbert, W.; Banin, E. Functional characterization of a novel "ulvan utilization loci" found in *Alteromonas* sp. LOR genome. *Algal Res.* **2017**, *25*, 39–46. [CrossRef]

17. Ulaganathan, T.; Helbert, W.; Kopel, M.; Banin, E.; Cygler, M. Structure–function analyses of a PL24 family ulvan lyase reveal key features and suggest its catalytic mechanism. *J. Biol. Chem.* **2018**, *293*, 4026–4036. [CrossRef]

18. Ulaganathan, T.; Boniecki, M.T.; Foran, E.; Buravenkov, V.; Mizrachi, N.; Banin, E.; Helbert, W.; Cygler, M. New ulvan-degrading polysaccharide lyase family: structure and catalytic mechanism suggests convergent evolution of active site architecture. *ACS Chem. Biol.* **2017**, *12*, 1269–1280. [CrossRef]

19. Ulaganathan, T.; Banin, E.; Helbert, W.; Cygler, M. Structural and functional characterization of PL28 family ulvan lyase NLR48 from *Nonlabens ulvanivorans*. *J. Biol. Chem.* **2018**, *293*, 11564–11573. [CrossRef]

20. Rydahl, M.G.; Kračun, S.K.; Fangel, J.U.; Michel, G.; Guillouzo, A.; Génicot, S.; Mravec, J.; Harholt, J.; Wilkens, C.; Motawia, M.S.; et al. Development of novel monoclonal antibodies against starch and ulvan-implications for antibody production against polysaccharides with limited immunogenicity. *Sci. Rep.* **2017**, *7*, 9326. [CrossRef]

21. Li, Y.; Li, W.; Zhang, G.; Lü, X.; Hwang, H.; Aker, W.G.; Wang, P. Purification and characterization of polysaccharides degradases produced by *Alteromonas* sp. A321. *Int. J. Biol. Macromol.* **2016**, *86*, 96–104. [CrossRef]

22. Qin, H.M.; Xu, P.; Guo, Q.; Cheng, X.; Gao, D.; Sun, D.; Zhu, Z.; Lu, F. Biochemical characterization of a novel ulvan lyase from *Pseudoalteromonas* sp. strain PLSV. *RSC Adv.* **2018**, *8*, 2610–2615. [CrossRef]

23. Reisky, L.; Stanetty, C.; Mihovilovic, M.D.; Schweder, T.; Hehemann, J.H.; Bornscheuer, U.T. Biochemical characterization of an ulvan lyase from the marine flavobacterium *Formosa agariphila* KMM 3901[T]. *Appl. Microbiol. Biotechnol.* **2018**, *102*, 6987–6996. [CrossRef] [PubMed]

24. Dickson, A.G. The measurement of sea water pH. *Mar. Chem.* **1993**, *44*, 131–142. [CrossRef]

25. Chen, L.; Feng, Y.; Zhou, Y.; Zhu, W.; Shen, X.; Chen, K.; Jiang, H.; Liu, D. Dual role of Zn^{2+} in maintaining structural integrity and suppressing deacetylase activity of SIRT1. *J. Inorg. Biochem.* **2010**, *104*, 180–185. [CrossRef]

26. Okino, S.; Ikeo, M.; Ueno, Y.; Taneda, D. Effects of Tween 80 on cellulase stability under agitated conditions. *Bioresour. Technol.* **2013**, *142*, 535–539. [CrossRef] [PubMed]

27. Yang, M.; Zhang, A.; Liu, B.; Li, W.; Xing, J. Improvement of cellulose conversion caused by the protection of Tween-80 on the adsorbed cellulase. *Biochem. Eng. J.* **2011**, *56*, 125–129. [CrossRef]

28. Xie, H.; Zhu, L.; Ma, T.; Wang, J.; Wang, J.; Su, J.; Shao, B. Immobilization of an enzyme from a *Fusarium* fungus WZ-I for chlorpyrifos degradation. *J. Environ. Sci.* **2010**, *22*, 1930–1935. [CrossRef]

29. Maruyama, Y.; Nakamichi, Y.; Itoh, T.; Mikami, B.; Hashimoto, W.; Murata, K. Substrate specificity of streptococcal unsaturated glucuronyl hydrolases for sulfated glycosaminoglycan. *J. Biol. Chem.* **2009**, *284*, 18059–18069. [CrossRef]

30. Lowry, O.H.; Rosebrough, N.J.; Farr, A.L.; Randall, R.J. Protein measurement with the Folin phenol reagent. *J. Biol. Chem.* **1951**, *193*, 265–275. [PubMed]

31. Laemmli, U.K. Cleavage of structural proteins during the assembly of the head of bacteriophage T4. *Nature* **1970**, *227*, 680–685. [CrossRef] [PubMed]

32. Qi, X.; Mao, W.; Gao, Y.; Chen, Y.; Chen, Y.; Zhao, C.; Li, N.; Wang, C.; Yan, M.; Lin, C.; et al. Chemical characteristic of an anticoagulant-active sulfated polysaccharide from *Enteromorpha clathrate*. *Carbohydr. Polym.* **2012**, *90*, 1804–1810. [CrossRef] [PubMed]

 *marine drugs* 

Article

Discovery of Two New Sorbicillinoids by Overexpression of the Global Regulator LaeA in a Marine-Derived Fungus *Penicillium dipodomyis* YJ-11

Jing Yu [1], Huan Han [1], Xianyan Zhang [1], Chuanteng Ma [1], Chunxiao Sun [1], Qian Che [1], Qianqun Gu [1], Tianjiao Zhu [1,2], Guojian Zhang [1,2] and Dehai Li [1,2,*]

[1] Key Laboratory of Marine Drugs, Chinese Ministry of Education, School of Medicine and Pharmacy, Ocean University of China, Qingdao 266003, China
[2] Laboratory for Marine Drugs and Bioproducts, Pilot National Laboratory for Marine Science and Technology (Qingdao), Qingdao 266237, China
* Correspondence: dehaili@ouc.edu.cn; Tel.: +86-532-8203-1619; Fax: +86-532-8203-3054

Received: 28 May 2019; Accepted: 26 July 2019; Published: 28 July 2019

Abstract: Overexpression of the global regulator LaeA in a marine-derived fungal strain of *Penicillium dipodomyis* YJ-11 induced obvious morphological changes and metabolic variations. Further chemical investigation of the mutant strain afforded a series of sorbicillinoids including two new ones named 10,11-dihydrobislongiquinolide (**1**) and 10,11,16,17-tetrahydrobislongiquinolide (**2**), as well as four known analogues, bislongiquinolide (**3**), 16,17-dihydrobislongiquinolide (**4**), sohirnone A (**5**), and 2′,3′-dihydrosorbicillin (**6**). The results support that the global regulator LaeA is a useful tool in activating silent gene clusters in *Penicillium* strains to obtain previously undiscovered compounds.

Keywords: genome mining; global regulator; LaeA; overexpression; *Penicillium dipodomyis*; sorbicillinoids

1. Introduction

Filamentous fungi have proven to be important sources of bioactive natural products for development of new drug leads. As the traditional approach (cultivation of microorganisms, chemical extraction of the produced metabolites, and final structure and bioactivity elucidations) has been continuously applied in discovery of new secondary metabolites, it has become a frequent issue that known structures are repeatedly discovered, while a big portion of biosynthetic genes are not expressed under current culturing technologies also termed as "silent" or "cryptic" genes. To increase the silent metabolic potential of the microbial producers, a variety of approaches such as heterologous expression, epigenetic regulation, transcriptional regulation and ribosome engineering, have been developed to affect the biosynthetic process in different gene regulation levels [1–3], among which, transcriptional factor regulation is often adopted because it is feasible and effective, and from which unexpected new secondary metabolites could be obtained by activation of the silent genes.

LaeA is an effective global regulator which was first discovered from *Aspergillus nidulans* and *A. fumigatus* by Jin Woo Bok and Nancy P. Keller in 2004 [4], and proved to be able to influence fungi in many aspects, such as increasing [5–7] or reducing secondary metabolite production [8], activating cryptic gene clusters [9], asexual and sexual differentiation as well as changes in phenotype including sporulation [5,9] and pigmentations [5]. Later on, LaeA gene analogues with similar functions were reported from other fungal species such as *Penicillium chrysogenum* [10], *Monascus ruber* [11], *Alternaria alternate* [12], and *Dothistroma septosporum* [8].

As a part of our ongoing work searching for diverse secondary metabolites from marine derived fungi, we recently isolated one filamentous fungi *Penicillium dipodomyis* YJ-11 from a marine sediment sample collected in Jiaozhou Bay of Qingdao. In order to activate the silent metabolic potential and obtain diversified secondary metabolites, we overexpressed the native global regulator of PdLaeA in *P. dipodomyis* YJ-11 and the mutant showed changes both in morphologies (sporulation and pigmentations) and metabolic profiles in contrast to the control (Figure 1). Further chemical studies on the mutant strain led to the isolation of two new compounds, 10,11-dihydrobislongiquinolide (**1**) and 10,11,16,17-tetrahydrobislongiquinolide (**2**), together with four known analogues, bislongiquinolide (**3**), also named as bisorbibutenolide, 16,17-dihydrobislongiquinolide (**4**), also named as dihydrotrichotetronine, sohirnone A (**5**), and 2′,3′-dihydrosorbicillin (**6**).

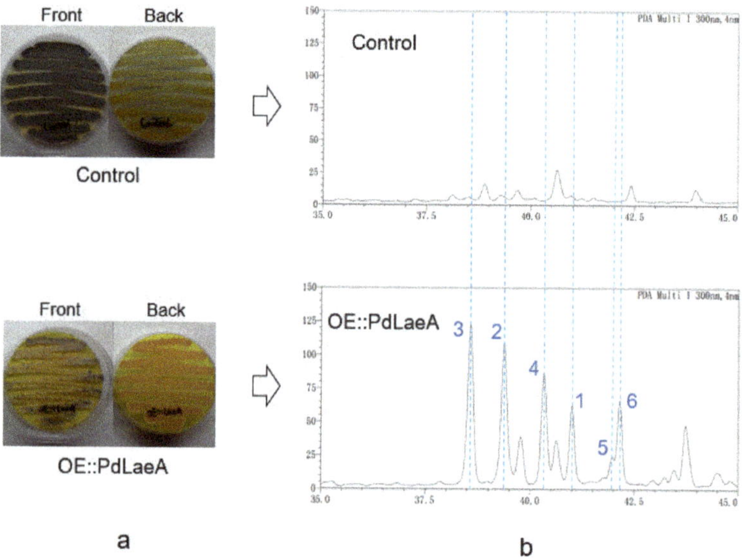

Figure 1. (**a**) Morphologies of the control strain and OE::PdLaeA strain of *P. dipodomyis* YJ-11 after incubating at 28 °C for 5 days for sporulation. (**b**) HPLC analysis of the extracts from the control strain and OE::PdLaeA strain of *P. dipodomyis* YJ-11.

2. Results and Discussion

The LaeA gene analogue PdLaeA was identified via Localblast, using a LaeA gene (*Aspergillus nidulans*, Q6TLK5.1) as a query. The PdLaeA gene was then cloned from genomic DNA of the strain *P. dipodomyis* YJ-11. The total size of the gene is 1283 bp and the predicted open reading frame (ORF) is 1053 bp, which may encode a polypeptide of 350 amino acids. BLAST analysis by NCBI indicated that the PdLaeA protein had 95% sequence identity to the protein of PcLaeA (*P. citrinum*, BAL61197.1), PrLaeA (*P. roqueforti FM164*, CDM34701.1), PdiLaeA (*P. digitatum PHI26*, EKV10385.1), and PdiLaeA (*P. digitatum Pd1*, XP_014530787.1). Phylogenetic analysis revealed that PdLaeA is mostly related to PcLaeA (Figure S2 in Supplementary Materials). Sequence analysis via InterProScan showed that the PdLaeA protein was an S-adenosyl-L-methionine-dependent methyltransferase (IPR029063), which is consistent with the putative mechanism of LaeA genes [13–15].

The PdLaeA in *P. dipodomyis* YJ-11 was amplified by specific primers (Table S1) and ligated into the vector pZeo using restriction sites XhoI and XbaI. The recombinant vector was transformed to *P. dipodomyis* YJ-11 to generate the OE::PdLaeA mutants (the mutant of *P. dipodomyis* YJ-11 harboring vacant pZeo was also generated as a control). The mutants showed obvious changes on both spore morphology and pigment formation. Followed by fermentation in PDB media with shaking at 28 °C for 9 days, high performance liquid chromatography (HPLC) analysis also showed a series of new peaks presenting in the extract of the OE::PdLaeA mutant compared with that of the control strain (Figure 1), indicating changes in secondary metabolite production.

For exploring the structures for the activation products, the OE::PdLaeA mutant was cultured in larger scale (10 L). Guided by UPLC-MS data, the EtOAc extract (12 g) of the fermentation was fractionated by octadecyl silane chemically bonded silica (ODS) medium performance liquid chromatography (MPLC) and then HPLC to yield compounds 10,11-dihydrobislongiquinolide (**1**, 11 mg), 10,11,16,17-tetrahydrobislongiquinolide (**2**, 35 mg), bislongiquinolide which also had been named as bisorbibutenolide [16,17] (**3**, 18 mg), 16,17-dihydrobislongiquinolide (also named as dihydrotrichotetronine [18,19], **4**, 84 mg), sohirnone A (**5**, 6 mg), and 2′,3′-dihydrosorbicillin (**6**, 5 mg) (Figure 2). The known compounds **3–6** were obtained and identified by the comparation of ^{1}H NMR data and mass spectroscopy with the literature refs. [16–22]. Compounds **3**, **5**, and **6** which had been isolated in our laboratory before [20], were also confirmed by HPLC analysis using standard samples.

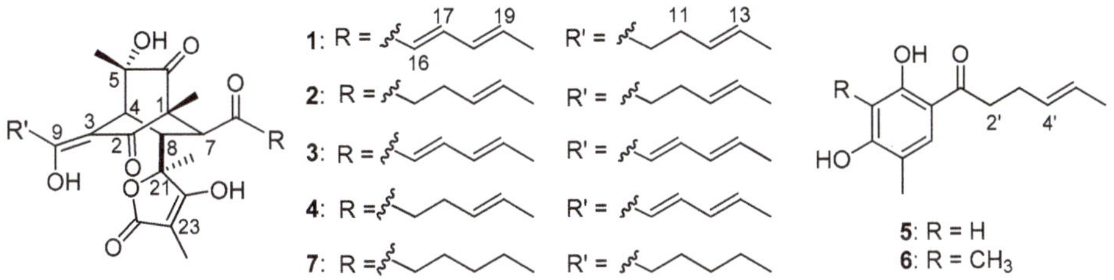

Figure 2. Structures of compounds **1–6** isolated from the strain OE::PdLaeA *Penicillium dipodomyis* YJ-11 and hydrogenation product **7**.

Compound **1** was obtained as a yellow powder with the molecular formula $C_{28}H_{34}O_8$ supported by the high resolution electrospray ionization mass spectroscopy (HRESIMS) peak at *m/z* 497.2185 [M − H]⁻ (calcd. 497.2170). The ^{1}H NMR spectrum of **1** revealed six methyl proton signals, two methylene proton signals, nine methine signals including six olefinic ones at 5.0–7.5 ppm, and three aliphatic ones at 2.8–3.5 ppm. In the ^{13}C NMR spectrum, in addition to the six methyls, two methylenes and nine methines which were assignable carbon signals, there were 11 quaternary carbons. The unsaturation degree further suggested the presence of three rings in the structure. The ^{1}H and ^{13}C NMR data (Table 1) were very similar to those of bislongiquinolide (**3**) [16,17] indicating that they had the same core structure. The major differences were the replacement of one double bond (in **3**) by a single bond (in **1**). Careful comparison of the ^{13}C NMR data of C_9-C_{14} (δ_C 181.3, 32.5, 28.4, 129.1, 126.9, 17.8 respectively) with those of compound **3** indicated the Δ^{10} double bond was reduced. The key homonuclear chemical shift correlation spectroscopy (COSY) correlations of H-10/H-11/H-12/H-13/H-14 together with ^{1}H detected heteronuclear multiple bond correlation (HMBC) correlations from H-11 (δ_H 2.42) to C-10 (δ_C 32.5), from H-10 (δ_H 2.51) to C-12 (δ_C 129.1), C-9 (δ_C 181.3), and C-3 (δ_C 108.4) (Figure 3) confirmed the structure of **1** as the 10,11-hydrogeneted analogue of **3**, which was named as 10,11-dihydrobislongiquinolide.

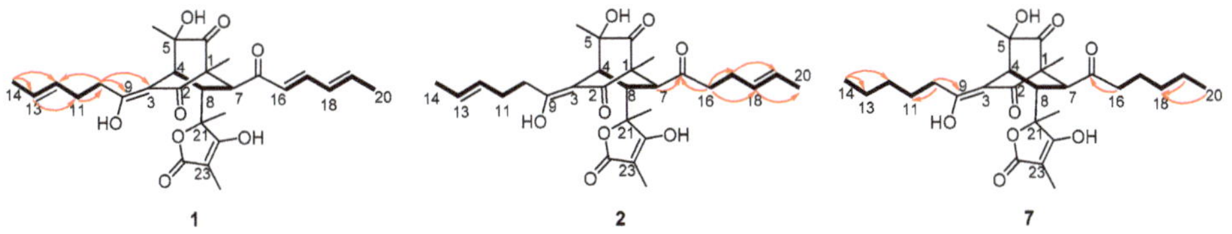

Figure 3. COSY and key HMBC correlations of compounds **1,2** and **7**.

Table 1. ^{1}H NMR data of experimental compounds **1** and **2** (500 MHz, CDCl$_3$, TMS, δ ppm), literature shared compound **3** [16] (400 MHz, CDCl$_3$, δ ppm), and the hydrogenation product **7** (600 MHz, CDCl$_3$, TMS, δ ppm) (*J* in Hz).

No.	1	2	3 [16]	7
1	–	–	–	–
2	–	–	–	–
3	–	–	–	–
4	3.36 s	3.42 s	3.36 br. s	3.39 s
5	–	–	–	–
6	–	–	–	–
7	2.95 d (7.0)	2.65 d (6.6)	3.43 d (4.8)	2.75 d (7.3)
8	3.20 br. d (7.0)	3.15 br. d (6.6)	3.21 dd (1.2, 4.8)	3.08 br. d (7.3)
9	–	–	–	–
10	2.51 t (7.4)	2.45-2.60 ^{a}	6.12 d (15.0)	2.40–2.50 ^{a}
11	2.42 m	2.39 m	7.33 dd (15.0, 10.6)	1.70 m
12	5.44 dt (15.3, 6.7)	5.44–5.55 ^{a}	6.28 dd (14.4, 10.6)	1.30–1.40 ^{a}
13	5.52 dq (15.3, 6.0)	5.44–5.55 ^{a}	6.24 dq (14.4, 6.8)	1.30–1.40 ^{a}
14	1.64 d (6.0)	1.59–1.67 ^{a}	1.88 d (6.8)	0.87 t (7.3)
15	–	–	–	–
16	6.02 d (15.0)	2.45–2.60 ^{a}	6.13 d (15.0)	2.40–2.50 ^{a}
17	7.16 dd (15.0, 10.6)	2.11, 2.19 m	7.22 dd (15.0, 10.4)	1.48 m
18	6.24 dd (15.0, 10.6)	5.34 m	6.21 dd (15.2, 10.4)	1.20–1.30 ^{a}
19	6.42 m	5.44–5.55 ^{a}	6.38 dq (15.2, 7.2)	1.20–1.30 ^{a}
20	1.93 d (6.7)	1.59–1.67 ^{a}	1.90 d (7.2)	0.89 t (6.5)
21	–	–	–	–
22	–	–	–	–
23	–	–	–	–
24	–	–	–	–
1-CH$_3$	1.05 s	1.13 s	1.12 s	1.11 s
5-CH$_3$	1.34 s	1.31 s	1.29 s	1.30 s
21-CH$_3$	1.57 s	1.55 s	1.51 s	1.50 s
23-CH$_3$	1.61 s	1.71 s	1.58 s	1.67 s
9-OH	14.45 br. s	14.45 br. s	13.99 br. s	14.46 br. s

^{a} Overlapped by other signals.

Compound **2** was obtained as a yellow oil and it was analyzed by HRESIMS (*m/z* 499.2343 [M − H]$^-$, calcd. 499.2326) for the molecular formula C$_{28}$H$_{36}$O$_8$, which was one double bond equivalent (DBE) less than **1**. The ^{1}H NMR spectrum of **2** revealed six methyl proton signals, four methylene proton signals, and seven methine protons with four olefinic ones (δ$_H$ 5.0–6.0) and three aliphatic ones (δ$_H$ 2.6–3.5). The ^{1}H and ^{13}C NMR data (Table 1) were very similar to those of **1** except that one double bond was hydrogenated in **2**. Careful comparison of the ^{13}C-NMR data of C$_{15}$-C$_{20}$ with those of compound **1** indicated that the C$_{16}$-C$_{17}$ double bond was reduced, which was further confirmed by the COSY correlations (H-16/H-17/H-18, H-19/H-20) and HMBC correlations from H-18 (δ$_H$ 5.34) to C-20 (δ$_C$ 17.8), from H-17 (δ$_H$ 2.11 and δ$_H$ 2.19) to C-19 (δ$_C$ 126.8), from H-16 (δ$_H$ 2.45–2.60) to C-15 (δ$_C$ 213.2), C-17 (δ$_C$ 25.9), and C-18 (δ$_C$ 128.5) (Figure 3). Thus, the structure of **2** was established and named as 10,11,16,17-tetrahydrobislongiquinolide.

The relative and absolute configurations of compounds **1** and **2** were determined by nuclear overhauser effect spectroscopy (NOESY) correlations (Figure 4), coupling constants (Table 1), hydrogenation reaction (Figure 5), circular dichroism (CD) spectra (Figure 6), and biogenetic considerations. In compound **1**, the *E* geometries of the double bonds in the side chains were deduced by the large coupling constants (15.0 Hz). In compound **2**, the NOESY correlation (Figure 4) between 1-CH$_3$ (δ$_H$ 1.13) and H-16 (δ$_H$ 2.45–2.60) suggested the same orientation of 1-CH$_3$ and the chain from C-7 through C-20; the NOESY correlation between H-4 (δ$_H$ 3.42) and H- 10 (δ$_H$ 2.45–2.60) indicated a *Z* geometry of Δ^{3,9} and located H-4 to the same side of the chain from C-9 through C-14, which was supported by the similar chemical shifts as reported for the known structure of bislongiquinolide [17] and bisorbibutenolide [16]. The NOESY correlation of 5-CH$_3$ (δ_H 1.31)/H-10 suggested that 5-CH$_3$ faced to C-3 in the bicyclo[2.2.2] ring. The 7R*, 8S* relative configuration was suggested by the coupling constant ($^3J_{\text{H-7, H-8}}$ = 6.6 Hz) [16]. The NOESY correlation of H-16/23-CH$_3$ (δ_H 1.71) indicated that the 23-CH$_3$ pointed to the direction of the side chain from C- 15 to C-20, while the NOESY correlations of H-10/21-CH$_3$ and H-4/21-CH$_3$ oriented 21-CH$_3$ to another side chain from C-9 to C-14 and H-4. Thus, the relative configuration of compound **2** should be 1R*, 4S*, 5S*, 7R*, 8S*, 21S*, which adopted the same configuration of the bicyclo[2.2.2]octanedione core as the reported analogues [16,17].

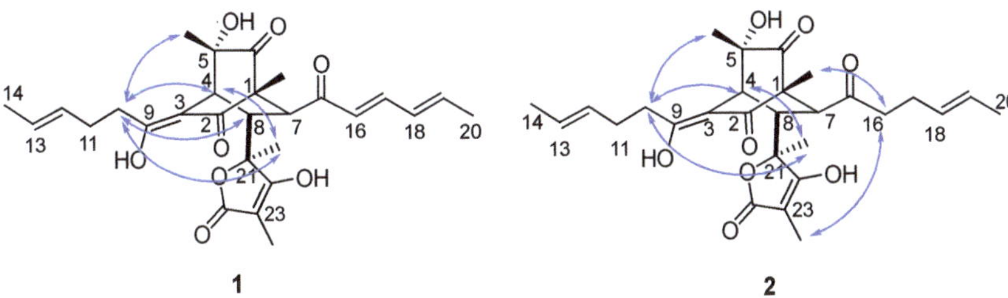

Figure 4. Key NOESY correlations of compounds **1** and **2**.

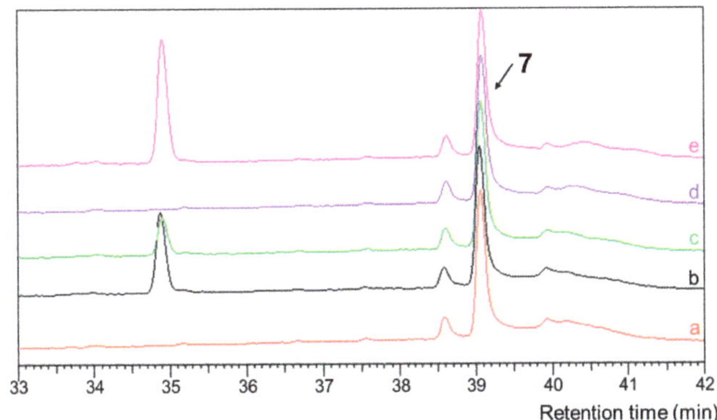

Figure 5. HPLC analysis of hydrogenation reaction products: (**a–d**) represent the reaction products of compounds **1–4** respectively and (**e**) is a co-injection of all the reaction products of compounds **1–4**.

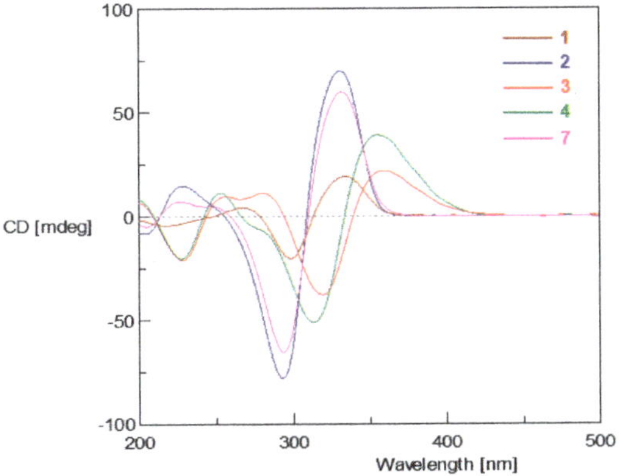

Figure 6. Circular dichroism (CD) spectra of **1–4** and **7**.

To confirm the consistent configurations of the co-isolated compounds **1–4**, hydrogenation over palladium on carbon was performed respectively. HPLC analysis of the reaction products showed that compounds **1–4** gave the same hydrogenation product (Figure 5), which suggested identical configurations of compounds **1–4** core structures. The hydrogenated product, named octahydrobislongiquinolide (**7**), was isolated by HPLC.

Compound **7** was obtained as a white powder with the molecular formula $C_{28}H_{40}O_8$ supported by the HRESIMS peak at m/z 527.2608 $[M + Na]^+$ (calcd. 527.2615). In addition to the signals expected for the bicyclo[2.2.2] core structure and the butyrolactone ring, the 1D NMR spectra of **7** shows signals for two methyls, eight methylenes, and no olefinic proton indicating that the double bonds in the sorbyl side chains were hydrogenated thoroughly. The key COSY correlations of H-10/H-11/H-12/H-13/H-14 together with HMBC correlations from H-14 (δ_H 0.87) to C-13 (δ_C 31.9) and C-12 (δ_C 22.5), from H-10 (δ_H 2.40–2.50) to C-11 (δ_C 25.2) and C-9 (δ_C 182.4) (Figure 3) confirmed the structure of one sidechain. Similarly, the key COSY correlations of H-16/H-17/H-18/H-19/H-20 together with HMBC correlations from H-16 (δ_H 2.40–2.50) to C-15 (δ_C 214.3), from H-19 (δ_H 1.20–1.30) to C-18 (δ_C 31.0), from H-20 (δ_H 0.89) to C-18 (δ_C 31.0) (Figure 3) confirm the structure of the other sidechain. Thus, the structure of **7** was established and named as octahydrobislongiquinolide.

Therefore, the absolute configurations of **1** and **2** were proposed to be the same as the co-isolated **3** and **4**, which were also supported by the optical rotation values (**1**, **2**, **3**, **4**: +35.6, +71.8, +105 [17], +350 [19]), the biogenetic consideration, and the similar trend in CD curves (Figure 6). Moreover, in the CD spectrum as shown in Figure 6, the curves of compounds **1–4** and **7** clustered into two groups (**1**, **2**, and **7** were in one group and **3** and **4** belonged to the other), which suggested that the double bond at C-10 on the 3-sorbyl substitution made a dominant contribution to the shape of the CD spectrum compared with the double bonds at other positions such as C-12, C-16 and C-18 in bicyclo[2.2.2]octanedione containing sorbicillinoid structures.

The cytotoxicities of compounds **1–4** were evaluated using HL-60, K562, BEL-7402, HCT-116, A549, Hela, L-02, MGC-803, SH-SY5Y, PC-3, H446, U87, MDA-MB-231, HO8910, ASPC-1 and MCF- 7 cell lines, but none of them presented a cytotoxic effect at 30 µM. The antimicrobial activity was also evaluated, with no activity detected under the concentration of 30 µM either. Compounds **1–4** exhibited identical weak siderophore activity with chrome azurol sulfonate (CAS) with an ED_{50} value of 400 µM (deferoxamine mesylate was used as a positive control with an ED_{50} value of 100 µM). Inspired by the radical scavenging activity of the known compound **3** [16], 1,1-diphenyl-2-picrylhydrazyl radical 2,2-diphenyl-1-(2,4,6-trinitrophenyl)hydrazyl (DPPH) radical scavenging assay was used to test the radical scavenging activity of compounds **2–4** (a paucity of compound **1** prevented analysis), and they showed similar effects with ED_{50} values of 166.7 µM, 183.3 µM, and 89.6 µM (the value of ascorbic acid was 27.3 µM as positive control).

The sorbicillinoids are a family of hexaketide metabolites, with the presence of the sorbyl sidechain as a unique structural feature. Monomeric, dimeric, trimeric, and nitrogen-containing metabolites constitute the sorbicillinoids, among which dimers are the most common, with potential bioactivities as drug leads [23,24]. The compounds **1–4** in this report are an expansion of the library of the bridged bicyclic bisorbicillinoids with the bicyclo[2.2.2]octanedione core structure. In contrast to the previous discovery strategies of conventional extraction and separation, biosynthesis or total synthesis, it is the first report of sorbicillinoid discovery by activating silent gene clusters. The above result shows that overexpression of the global regulator LaeA is a useful method to discover new natural products by activating silent biosynthetic gene clusters.

3. Materials and Methods

3.1. General Experimental Procedures

DNA restriction enzymes were used as recommended by the manufacturer (New England Biolabs, NEB, Beijing, China). Polymerase chain reaction (PCR) was performed using *TransStart®* *Fastpfu* DNA

Polymerase (Transgen Biotech, Beijing, China). PCR products were confirmed by PCR analysis using 2×*EasyTaq*® *PCR SuperMix* (Transgen Biotech, Beijing, China). Genomic DNA samples were prepared using the CTAB isolation buffer at pH 8.0 (20 g/L cetyltrimethylammonium bromide, 1.4 M sodium chloride, and 20 mM EDTA) [25]. The gene-specific primers are listed in Table S1. UV spectra were recorded on a Beckman DU 640 spectrophotometer (Beckman Coulter Inc., Brea, CA, USA). Specific optical rotations were obtained using a JASCO P-1020 digital polarimeter (JASCO Corporation, Tokyo, Japan). Electrospray ionization-mass spectrometry (ESIMS) and HRESIMS were obtained on a Thermo Scientific LTQ Orbitrap XL mass spectrometer (Thermo Fisher Scientific, Waltham, MA, USA) or using a Micromass Q-TOF ULTIMA GLOBAL GAA076 LC Mass spectrometer (Wasters Corporation, Milford, MA, USA). ^{1}H NMR, ^{13}C NMR and 2D NMR spectra were recorded on an Agilent 500 MHz DD2 spectrometer (Agilent Technologies Inc., Santa Clara, CA, USA). LC-MS was performed using an Acquity UPLC H-Class coupled to a SQ Detector 2 mass spectrometer using a BEH C18 column (1.7 μm, 2.1 × 50 mm, 1 mL/min) (Waters Corporation, Milford, MA, USA). Semi-preparative HPLC (YMC Co., Ltd., Kyoto, Japan) was performed on an ODS column (YMC-Pack ODS-A, 10 × 250 mm, 5 μm, 3 mL/min). Medium-pressure liquid chromatography (MPLC) was performed on a Bona-Agela CHEETAHTM HP100 (Beijing Agela Technologies Co., Ltd., Beijing, China) [26].

3.2. Materials and Culture Conditions

The fungal strain, authenticated as *P. dipodomyis* YJ-11, was collected from the marine sediment around the sewage outlet in Jiaozhou Bay of Qingdao. The strain was identified by internal transcribed spacer (ITS) sequence and the sequence data were submitted to GenBank (accession number: MK682303). Working stocks were prepared on Potato Dextrose agar slants stored at 4 °C in our laboratory.

The strain was incubated in media potato dextrose agar (PDA, 20% potato, 2% dextrose, and 2% agar) at 28 °C for 5 days for sporulation. Media potato sorbitol agar (PSA, 20% potato, 2% dextrose, 1.2 M D-sorbitol, and 2% agar) and PDA with 400 μg/mL zeocin (Sigma) were used to screen resistant transformants. The mutants were also cultured at 28 °C in an incubator. For compound production, the strains were cultured on potato dextrose (PDB, 20% potato and 2% dextrose) at 28 °C, 180 rpm for 9 days. *Trans*1-T1 Phage Resistant Chemically competent cell (Transgen Biotech, Beijing, China) was used for plasmid preservation and amplification, following standard recombinant DNA techniques. *E. coli* cultures were growing at 37 °C in another incubator.

3.3. Sequence Analysis of the PdLaeA Gene

The LaeA gene was analyzed by Localblast with the reported LaeA obtained in national center for biotechnology information (NCBI). For the multiple sequence alignment analysis, the amino acid sequences of PdLaeA and other LaeA homologues from different species retrieved from NCBI were aligned using the ClustalX software [27]. The phylogenetic analysis was conducted with the MEGA7 software [28]. The conserved domain of the PdLaeA protein was scanned by the InterProScan program [29].

3.4. Construction of the PdLaeA Expression Vector

The overexpression vector pZeo which mainly contains a constitutive promoter PgpdA, resistant ampicillin gene and resistant zeocin gene as selection markers was digested with restriction endonucleases XhoI and XbaI. The PdLaeA gene was PCR-amplified (from genomic DNA of the wild-type strain *P. dipodomyis* YJ-11) using specific primers containing XhoI and XbaI restriction sites (Table S1). The PCR products were digested with the same endonucleases and PdLaeA gene was introduced into pZeo vector to generate pZeo-PdLaeA (Figure S3 in Supplementary Materials). The recombinant vector was transformed into competent *E. coli* strain *Trans*1-T1 to extract plasmids for transformation.

3.5. Fungal Protoplast Formation and Transformation

The strain *P. dipodomyis* YJ-11 was first grown on PDA plates at 28 °C for 5 days. Fresh spores were collected into 50 mL PDB together with yeast extract media in 250 mL Erlenmeyer flasks and germinated at 28 °C and 180 rpm for about 7 h. Mycelia were gathered by centrifugation at 4000 rpm for 15 min, and washed by 25 mL osmotic buffer (1.2 M $MgSO_4$, 10 mM sodium phosphate, pH 5.8). Subsequently, the mycelia were suspended into 10 mL of osmotic buffer containing 30 mg lysing enzymes from *Trichodema harzianum* (Sigma) and 20 mg Yatalase (TaKaRa), transferred into an empty sterile bottle, and cultured in a shaker of 28 °C at 80 rpm overnight to form protoplast.

After the whole night, the mixture was collected in a 50 mL centrifuge tube and covered gently with isopyknic protoplast trapping buffer (0.6 M sorbitol, 0.1 M pH 7.0 Tris-HCl). After centrifugation at 4000 rpm for 15 min at 4 °C, protoplasts were collected in the interface of the above two buffers. The protoplasts were then transferred to a sterile 50 mL centrifuge tube and washed by 20 mL STC buffer (1.2 M sorbitol, 10 mM $CaCl2$, 10 mM pH 7.5 Tris-HCl). The protoplasts were resuspended in 2 mL STC buffer for transformation. Then, the freeze-drying plasmids pZeo-PdLaeA and the plasmid pZeo (the desired mutants were regarded as a control in the following crude analysis) were dissolved in 50 μL STC buffer, followed by 100 μL protoplast suspension and the mixture incubated for 60 min on ice. Next, 600 μL of polyethylene glycol (PEG) solution (60% PEG, 50 mM calcium chloride and 50 mM pH 7.5 Tris-HCl) was added to the protoplast mixture, and the mixture was incubated at room temperature for an additional 25 min. The mixture was spread on the regeneration solid PSA medium (PDA medium with 1.2 M sorbitol and 400 μg/mL zeocin) and incubated at 28 °C for around 4 days [30].

3.6. Transformants Screening

After regeneration, the transformants were passaged to PDA plates with 400 μg/mL zeocin respectively. Zeocin resistant mutants were transferred onto new PDA media containing 400 μg/mL zeocin for the second screening. The strains that were able to grow were subjected to further PCR analysis validation. The putative OE::PdLaeA mutants and the wild-type strain were cultured on PDA media for 5 days at 28 °C in an incubator in order to extract genomic DNAs. PCR analysis to confirm the gene insertion was carried out using three pairs of primers to verify the zeocin resistant mutants as shown in Table S1 and Figure S4 (primers gpda-1 and gpda-2 to verify whole of the PdLaeA gene, primers gpda-1 and YZ-LaeA-F to verify the upstream of target gene, primers gpda-2 and YZ-LaeA-R to verify the downstream of target gene). Those transformants which went through all these verifications were recognized as the desired mutants.

3.7. Fermentation and Extraction

For small-scale analysis, the OE::PdLaeA mutant strains and the control strain were grown on PDA plates for 5 days at 28 °C. Shortly after sporing, they were inoculated into 150 mL of PDB medium and cultured at 28 °C, 180 rpm. Nine days later, the cultures were extracted with twice the volume of ethyl acetate. The organic phase was evaporated and the residue was dissolved in MeOH, which was analyzed by HPLC and indicated that one of the six mutants showed an apparent change in metabolite production (Figure 1).

For compound isolation, the selected strain was initially handled the same as above. Then a large-scale fermentation was performed in 500 mL Erlenmeyer flasks (total 10 L) for further incubation. The broth was extracted three times with ethyl acetate to give a total of 60 L of extract solution. The organic phase was evaporated under reduced pressure to afford a crude residue (12 g).

3.8. Compound Isolation

The crude extracts were separated by MPLC (60% to 100% MeOH in H_2O for 60 min). The fractions containing the target compounds were combined and concentrated to give 7 fractions (Fr.1 to Fr.7). For further purification, semi-preparative HPLC were carried out. Fr.4 was then purified by HPLC

(55% MeCN in H_2O, 0.5% THF) to obtain Fr.4-1 (**6**; 5 mg; t_R 24.2 min). Fr.5 was purified by HPLC (42% MeCN in H_2O, 0.5% THF) to obtain Fr.5-1 (**3**;18 mg; t_R 30.8 min) and Fr.5-2 (**1**; 7 mg; t_R 38.0 min). Fr.6 was purified by HPLC (60% MeOH in H_2O, 0.5% THF) to give five subfractions (Fr.6- 1 to Fr.6-5), in which Fr.6-1 (**1**; 4 mg; t_R 30.3 min), Fr.6-4 (**2**; 35 mg; t_R 46.1 min), Fr.6-5 (**5**; 6 mg; t_R 52.2 min) were proven to be pure. Fr.6-3-1 (**4**; 84 mg; t_R 20.5 min) was purified by HPLC (50% MeCN in H_2O, 0.5% THF) from Fr.6-3. The purity of each compound was checked by LC-MS and the structures were confirmed by NMR including ^{1}H, ^{13}C, and 2D NMR spectra.

10,11-Dihydrobislongiquinolide (**1**): $[\alpha]_D^{20}$ 35.6 (c 0.13, MeOH), UV (MeOH) λ_{max} (log ε): 228 (3.67), 296 (2.97) nm, ^{1}H-NMR (CDCl$_3$, 500 MHz) and ^{13}C-NMR (CDCl$_3$, 125 MHz) data are shown in Tables 1 and 2, HRESIMS *m/z*: 497.2185 [M − H]$^-$ (Calcd. for $C_{28}H_{34}O_8$: 497.2170).

10,11,16,17-Tetrahydrobislongiquinolide (**2**): $[\alpha]_D^{20}$ 71.8 (c 0.50, MeOH), UV (MeOH) λ_{max} (log ε): 237 (1.22), 296 (3.24) nm, ^{1}H-NMR (CDCl$_3$, 500 MHz) and ^{13}C-NMR (CDCl$_3$, 125 MHz) data are shown in Tables 1 and 2, HRESIMS *m/z*: 499.2343 [M − H]$^-$ (Calcd. for $C_{28}H_{36}O_8$: 499.2326).

Table 2. ^{13}C NMR data of experimental compounds **1** and **2** (125 MHz, CDCl$_3$, TMS, δ ppm), literature shared compound **3** [16] (100 MHz, CDCl$_3$, δ ppm) and the hydrogenation product **7** (150 MHz, CDCl$_3$, TMS, δ ppm).

No.	1	2	3 [16]	7
1	61.7	61.4	62.6	61.4
2	193.0	192.9	194.9	192.9
3	108.4	108.4	108.5	106.3
4	42.8	42.7	42.4	42.9
5	75.3	75.2	75.0	75.3
6	209.9	209.1	208.3	209.2 ^{b}
7	50.7	54.0	51.3	54.3
8	42.0	42.6	43.6	42.7
9	181.3	181.2	169.8	182.4
10	32.5	32.4	117.6	32.4
11	28.4	28.4	143.9	25.2
12	129.1	129.0	130.3	22.5
13	126.9	126.9	140.9	31.9
14	17.8	17.8	18.9	14.0
15	202.1	213.2	202.7	214.3
16	127.2	45.9	127.0	46.0
17	148.3	25.9	148.0	22.5
18	130.1	128.5	131.0	31.0
19	146.0	126.8	145.5	29.8
20	19.2	17.8	19.1	14.0
21	83.1	84.0	83.2	83.1
22	175.8	177.3	176.5	176.0
23	98.8	97.8	98.2	98.2
24	174.4	175.4	174.9	173.6
1-CH$_3$	10.5	10.6	11.0	10.7
5-CH$_3$	24.1	24.0	23.5	24.2
21-CH$_3$	21.8	21.3	23.1	21.5
23-CH$_3$	6.1	5.9	6.3	6.1

^{b} Detected by HMBC spectrum.

3.9. Hydrogenation of Compounds 1–4

Compounds **1–4** in methanol were hydrogenated over palladium on carbon at room temperature overnight, respectively. The reaction mixture was filtered, concentrated, and purified by semi-preparative HPLC (65% MeOH in H_2O, 0.1% THF) to give octahydrobislongiquinolide (**7**; 5 mg;

t_R 15.6 min). The purity of compound **7** was checked by LC-MS and the structures were confirmed by NMR including ^{1}H, ^{13}C and 2D NMR spectra.

Octahydrobislongiquinolide (**7**): $[\alpha]_D^{20}$ 78.4 (c 0.10, MeOH), UV (MeOH) λ_{max} (log ε): 227 (4.78), 297 (2.13) nm, ^{1}H-NMR (CDCl$_3$, 600 MHz) and ^{13}C-NMR (CDCl$_3$, 150 MHz) data are shown in Tables 1 and 2, HRESIMS *m/z*: 527.2608 [M + Na]$^+$ (calcd. 527.2615 for $C_{28}H_{40}O_8Na$).

3.10. Assay of Cytotoxicity, Antimicrobial and Antioxidation Activity

These biological evaluations were carried out as previously reported in References [26,31].

3.11. Assay of Siderophore Activity

Siderophore activity was evaluated by the chrome azurol sulfonate (CAS) assay. Buffer A consist of 50 mL 5 M CAS solution and 10 mL Fe^{3+} solution (1 mM FeCl$_3 \cdot 6H_2O$, 10 mM HCl). 72.9 mg hexadecyl trimethyl ammonium bromide (HDTMA) was dissolved in 40 mL water after heated as buffer B. Buffer B was then slowly added to buffer A under stirring to afford 100 mL CAS assay solution. All tested compounds were dissolved in MeOH at stepwise concentrations (1–100 mM) then 99 µL of the compound solution and 1 µL CAS assay solution were dispensed into wells of a 96-well microtiter tray. The mixture was shaken and left to stand for 4 h. After that, absorbance was measured at 630 nm and the inhibition rate was calculated. Deferoxamine mesylate, which could make the color change of CAS from blue to orange, was used to determine standard curves relating the CAS reactivity to iron-binding ligands. The ED$_{50}$ values denoted the concentration of sample required to remove 50% of the iron in CAS solution [32,33].

4. Conclusions

In summary, guided by bioinformatic analysis of the genomic sequence of *P. dipodomyis* YJ-11, we discovered one global regulator of PdLaeA. Further overexpression of the PdLaeA gene in *P. dipodomyis* YJ-11 led to the discovery of two new compounds, 10,11-dihydrobislongiquinolide (**1**), 10,11,16,17-tetrahydrobislongiquinolide (**2**), together with four known sorbicillin analogues (compounds **3–6**). This is the first report on application of global regulator LaeA in *P. dipodomyis* with the purpose of increasing the secondary metabolite producing potential. This result also indicated that the production of sorbicillinoids may be regulated by the global regulator LaeA.

Supplementary Materials: The following are available online at http://www.mdpi.com/1660-3397/17/8/446/s1, Figure S1: AntiSMASH analysis of the genome of the strain *Penicillium dipodomyis* YJ-11, Figure S2: Phylogenetic tree analysis of PdLaeA and its homologs from different species, Figure S3: Map of the vector pZeo and PdLaeA overexpression plasmid pZeo-PdLaeA, Figure S4: PCR analysis for confirming the gene insertion, Figures S5–S11: 1D, 2D NMR and HRESIMS spectra of 10,11-dihydrobislongiquinolide (**1**), Figures S12–S18: 1D, 2D NMR and HRESIMS spectra of 10,11,16,17-tetrahydrobislongiquinolide (**2**). Figures S19–S20: 1D NMR spectra of bislongiquinolide (**3**). Figures S21–S22: 1D NMR spectra of 16,17-dihydrobislongiquinolide (**4**), Figures S23–S28: 1D, 2D NMR and HRESIMS spectra of octahydrobislongiquinolide (**7**), Table S1: The primers used in this study.

Author Contributions: The contributions of the respective authors are as follows: J.Y. drafted the work, constructed the plasmids and performed the fermentation, extraction, as well as the isolation. X.Z., H.H., and C.M. were involved in the acquisition of mutant strains and bioinformatic analysis. J.Y. and C.S. elucidated the constituents and performed the biological evaluations. Q.C., Q.G., T.Z., and G.Z. contributed to checking and confirming all procedures of isolation and identification. D.L. designed the study, supervised the laboratory work, and contributed to the critical reading of the manuscript, and was also involved in structural determination and bioactivity elucidation. All the authors have read the final manuscript and approved the submission.

Funding: This work was financially supported by the Fundamental Research Funds for the Central Universities (201941001), NSFC-Shandong Joint Fund for Marine Science Research Centers (U1606403), Financially supported by the Marine S&T Fund of Shandong Province for Pilot National Laboratory for Marine Science and Technology (Qingdao) (2018SDKJ0401-2), Taishan Scholar Youth Expert Program in Shandong Province (tsqn201812021).

Acknowledgments: The authors recognize material contributions from Prof. Yi Tang (University of California, Los Angeles) for providing plasmids.

Conflicts of Interest: The authors declare no conflict of interest.

References

1. Brakhage, A.A.; Schuemann, J.; Bergmann, S.; Scherlach, K.; Schroeckh, V.; Hertweck, C. Activation of fungal silent gene clusters: A new avenue to drug discovery. *Prog. Drug Res.* **2008**, *66*, 3–12. [CrossRef]
2. Netzker, T.; Fischer, J.; Weber, J.; Mattern, D.J.; König, C.C.; Valiante, V.; Schroeckh, V.; Brakhage, A.A. Microbial communication leading to the activation of silent fungal secondary metabolite gene clusters. *Front. Microbiol.* **2015**, *6*, 299. [CrossRef] [PubMed]
3. Rutledge, P.J.; Challis, G.L. Discovery of microbial natural products by activation of silent biosynthetic gene clusters. *Nat. Rev. Microbiol.* **2015**, *13*, 509–523. [CrossRef] [PubMed]
4. Bok, J.W.; Keller, N.P. LaeA, a Regulator of Secondary Metabolism in *Aspergillus* spp. *Eukaryot. Cell* **2004**, *3*, 527–535. [CrossRef] [PubMed]
5. Lee, S.S.; Lee, J.H.; Lee, I. Strain improvement by overexpression of the laeA gene in *Monascus pilosus* for the production of monascus-fermented rice. *J. Microbiol. Biotechnol.* **2013**, *23*, 959–965. [CrossRef] [PubMed]
6. Hong, E.J.; Kim, N.K.; Lee, D.; Kim, W.G.; Lee, I. Overexpression of the laeA gene leads to increased production of cyclopiazonic acid in *Aspergillus fumisynnematus*. *Fungal Biol.* **2015**, *119*, 973–983. [CrossRef] [PubMed]
7. Linde, T.; Zoglowek, M.; Lubeck, M.; Frisvad, J.C.; Lubeck, P.S. The global regulator LaeA controls production of citric acid and endoglucanases in *Aspergillus carbonarius*. *J. Ind. Microbiol. Biot.* **2016**, *43*, 1139–1147. [CrossRef] [PubMed]
8. Chettri, P.; Bradshaw, R.E. LaeA negatively regulates dothistromin production in the pine needle pathogen *Dothistroma septosporum*. *Fungal Genet. Biol.* **2016**, *97*, 24–32. [CrossRef] [PubMed]
9. Jiang, T.; Wang, M.; Li, L.; Si, J.; Song, B.; Zhou, C.; Yu, M.; Wang, X.; Zhang, Y.; Ding, G.; et al. Overexpression of the Global Regulator LaeA in *Chaetomium globosum* Leads to the Biosynthesis of Chaetoglobosin, Z.J. *Nat. Prod.* **2016**, *79*, 2487–2494. [CrossRef]
10. Kosalkova, K.; Garcia-Estrada, C.; Ullan, R.V.; Godio, R.P.; Feltrer, R.; Teijeira, F.; Mauriz, E.; Martin, J.F. The global regulator LaeA controls penicillin biosynthesis, pigmentation and sporulation, but not roquefortine C synthesis in *Penicillium chrysogenum*. *Biochimie* **2009**, *91*, 214–225. [CrossRef]
11. Liu, Q.; Cai, L.; Shao, Y.; Zhou, Y.; Li, M.; Wang, X.; Chen, F. Inactivation of the global regulator LaeA in *Monascus ruber* results in a species-dependent response in sporulation and secondary metabolism. *Fungal Biol.* **2016**, *120*, 297–305. [CrossRef] [PubMed]
12. Estiarte, N.; Lawrence, C.B.; Sanchis, V.; Ramos, A.J.; Crespo-Sempere, A. LaeA and VeA are involved in growth morphology, asexual development, and mycotoxin production in *Alternaria alternata*. *Int. J. Food Microbiol.* **2016**, *238*, 153–164. [CrossRef] [PubMed]
13. Keller, N.P.; Turner, G.; Bennett, J.W. Fungal secondary metabolism—From biochemistry to genomics. *Nat. Rev. Microbiol.* **2005**, *3*, 937–947. [CrossRef] [PubMed]
14. Bok, J.W.; Noordermeer, D.; Kale, S.P.; Keller, N.P. Secondary metabolic gene cluster silencing in *Aspergillus nidulans*. *Mol. Microbiol.* **2006**, *61*, 1636–1645. [CrossRef] [PubMed]
15. Xu, Z.C.; Sun, C.; Xu, J.; Zhang, X.; Luo, H.M.; Ji, A.J.; Hu, Y.L.; Song, J.Y.; Chen, S.L. Progress in the study of Velvet and LaeA proteins and their relation to the development and bioactive compounds in medicinal fungi. *Yao Xue Xue Bao* **2014**, *49*, 1520–1527. [PubMed]
16. Abe, N.; Murata, T.; Hirota, A. Novel oxidized sorbicillin dimers with 1,1-diphenyl-2-picrylhydrazyl-radical scavenging activity from a fungus. *Biosci. Biotechnol. Biochem.* **1998**, *62*, 2120–2126. [CrossRef] [PubMed]
17. Andrade, R.; Ayer, W.A.; Trifonov, L.S. The metabolites of *Trichoderma Longibrachiatum*. III. Two new tetronic acids: 5-hydroxyvertinolide and bislongiquinolide. *Aust. J. Chem.* **1997**, *50*, 255–257. [CrossRef]
18. Shirota, O.; Pathak, V.; Hossain, C.F.; Sekita, S.; Takatori, K.; Satake, M. Structural elucidation of trichotetronines: Polyketides possessing a bicyclo[2.2.2]octane skeleton with tetronic acid moiety isolated from *Trichoderma* sp. *J. Chem. Soc. Perkin Trans.* **1997**, *29*, 2961–2964. [CrossRef]
19. Evidente, A.; Andolfi, A.; Cimmino, A.; Ganassi, S.; Altomare, C.; Favilla, M.; Cristofaro, A.D.; Vitagliano, S.; Sabatini, M.A. Bisorbicillinoids Produced by the Fungus *Trichoderma Citrinoviride* Affect Feeding Preference of the Aphid *Schizaphis Graminum*. *J. Chem. Ecol.* **2009**, *35*, 533–541. [CrossRef]

20. Du, L.; Zhu, T.; Li, L.; Cai, S.; Zhao, B.; Gu, Q. Cytotoxic sorbicillinoids and bisorbicillinoids from a marine-derived fungus *Trichoderma* sp. *Chem. Pharm. Bull.* **2009**, *57*, 220–223. [CrossRef]

21. Trifonov, L.S.; Bieri, J.H.; Prewo, R.; Dreiding, A.S.; Hoesch, L.; Rast, D.M. Isolation and structure elucidation of three metabolites from *Verticillium intertextum*: Sorbicillin, dihydrosorbicillin and bisvertinoquinol. *Tetrahedron* **1983**, *39*, 4243–4256. [CrossRef]

22. Maskey, R.P.; Grün-Wollny, I.; Laatsch, H. Sorbicillin analogues and related dimeric compounds from *Penicillium Notatum*. *J. Nat. Prod.* **2005**, *68*, 865–870. [CrossRef]

23. Balde, E.S.; Andolfi, A.; Bruyère, C.; Cimmino, A.; Lamoral-Theys, D.; Vurro, M.; Damme, M.V.; Altomare, C.; Mathieu, V.; Kiss, R.; et al. Investigations of fungal secondary metabolites with potential anticancer activity. *J. Nat. Prod.* **2010**, *73*, 969. [CrossRef]

24. Harned, A.M.; Volp, K.A. The sorbicillinoid family of natural products: Isolation, biosynthesis, and synthetic studies. *Nat. Prod. Rep.* **2011**, *28*, 1790. [CrossRef]

25. Tang, M.C.; Cui, X.; He, X.; Ding, Z.; Zhu, T.; Tang, Y.; Li, D. Late-Stage Terpene Cyclization by an Integral Membrane Cyclase in the Biosynthesis of Isoprenoid Epoxycyclohexenone Natural Products. *Org. Lett.* **2017**, *19*, 5376–5379. [CrossRef]

26. Yu, G.; Wu, G.; Sun, Z.; Zhang, X.; Che, Q.; Gu, Q.; Zhu, T.; Li, D.; Zhang, G. Cytotoxic Tetrahydroxanthone Dimers from the Mangrove-Associated Fungus *Aspergillus versicolor* HDN1009. *Mar. Drugs* **2018**, *16*, 335. [CrossRef]

27. Larkin, M.A.; Blackshields, G.; Brown, N.P.; Chenna, R.; McGettigan, P.A.; McWilliam, H.; Valentin, F.; Wallace, I.M.; Wilm, A.; Lopez, R.; et al. Clustal W and Clustal X version 2.0. *Bioinformatics* **2007**, *23*, 2947–2948. [CrossRef]

28. Kumar, S.; Stecher, G.; Tamura, K. Molecular Evolutionary Genetics Analysis Version 7.0 for bigger datasets. *Mol. Biol. Evol.* **2016**, *33*, 1870–1874. [CrossRef]

29. Mitchell, A.L.; Attwood, T.K.; Babbitt, P.C.; Blum, M.; Bork, P.; Bridge, A.; Brown, S.D.; Chang, H.Y.; El-Gebali, S.; Fraser, M.I.; et al. InterPro in 2019: Improving coverage, classification and access to protein sequence annotations. *Nucleic Acids Res.* **2019**, *47*, 351–360. [CrossRef]

30. Ohashi, M.; Liu, F.; Hai, Y.; Chen, M.; Tang, M.C.; Yang, Z.; Sato, M.; Watanabe, K.; Houk, K.N.; Tang, Y. SAM-dependent enzyme-catalysed pericyclic reactions in natural product biosynthesis. *Nature* **2017**, *549*, 502–506. [CrossRef]

31. Zhang, Z.; He, X.; Che, Q.; Zhang, G.; Zhu, T.; Gu, Q.; Li, D. Sorbicillasins A-B and Scirpyrone K from a Deep-Sea-Derived Fungus, *Phialocephala* sp. FL30r. *Mar. Drugs* **2018**, *16*, 245. [CrossRef]

32. Schwyn, B.; Neilands, J.B. Universal chemical assay for the detection and determination of siderophores. *Anal. Biochem.* **1987**, *160*, 47–60. [CrossRef]

33. Jang, J.H.; Kanoh, K.; Adachi, K.; Matsuda, S.; Shizuri, Y. Tenacibactins a-d, hydroxamate siderophores from a marine-derived bacterium, *Tenacibaculum* sp. a4k-17. *J. Nat. Prod.* **2007**, *70*, 563–566. [CrossRef]

Article

Genome Sequencing and Analysis of *Thraustochytriidae* sp. SZU445 Provides Novel Insights into the Polyunsaturated Fatty Acid Biosynthesis Pathway

Xingyu Zhu [1,†], Shuangfei Li [1,2,3,†], Liangxu Liu [1], Siting Li [1], Yanqing Luo [1], Chuhan Lv [1], Boyu Wang [1,2,3], Christopher H. K. Cheng [4], Huapu Chen [5] and Xuewei Yang [1,2,3,*]

[1] Guangdong Technology Research Center for Marine Algal Bioengineering, Guangdong Key Laboratory of Plant Epigenetics, College of Life Sciences and Oceanography, Shenzhen University, Shenzhen 518060, China; staruny9562@163.com (X.Z.); sfli@szu.edu.cn (S.L.); liangxuliu@icloud.com (L.L.); siting.li@aut.ac.nz (S.L.); lyanqingl87@163.com (Y.L.); LVCHU0510@163.com (C.L.); 17612760955@163.com (B.W.)

[2] Shenzhen Key Laboratory of Marine Biological Resources and Ecology Environment, Shenzhen Key Laboratory of Microbial Genetic Engineering, College of Life Sciences and Oceanography, Shenzhen University, Shenzhen 518055, China

[3] Longhua Innovation Institute for Biotechnology, Shenzhen University, Shenzhen 518060, China

[4] School of Biomedical Sciences, The Chinese University of Hong Kong, Hong Kong 999077, China; chkcheng@cuhk.edu.hk

[5] Guangdong Research Center on Reproductive Control and Breeding Technology of Indigenous Valuable Fish Species, Fisheries College, Guangdong Ocean University, Zhanjiang 524088, China; chpsysu@hotmail.com

* Correspondence: yangxw@szu.edu.cn

† Co-first author.

Received: 23 December 2019; Accepted: 14 February 2020; Published: 18 February 2020

Abstract: *Thraustochytriidae* sp. have broadly gained attention as a prospective resource for the production of omega-3 fatty acids production in significant quantities. In this study, the whole genome of *Thraustochytriidae* sp. SZU445, which produces high levels of docosapentaenoic acid (DPA) and docosahexaenoic acid (DHA), was sequenced and subjected to protein annotation. The obtained clean reads (63.55 Mb in total) were assembled into 54 contigs and 25 scaffolds, with maximum and minimum lengths of 400 and 0.0054 Mb, respectively. A total of 3513 genes (24.84%) were identified, which could be classified into six pathways and 44 pathway groups, of which 68 genes (1.93%) were involved in lipid metabolism. In the Gene Ontology database, 22,436 genes were annotated as cellular component (8579 genes, 38.24%), molecular function (5236 genes, 23.34%), and biological process (8621 genes, 38.42%). Four enzymes corresponding to the classic fatty acid synthase (FAS) pathway and three enzymes corresponding to the classic polyketide synthase (PKS) pathway were identified in *Thraustochytriidae* sp. SZU445. Although PKS pathway-associated dehydratase and isomerase enzymes were not detected in *Thraustochytriidae* sp. SZU445, a putative DHA- and DPA-specific fatty acid pathway was identified.

Keywords: whole-genome sequencing; docosahexaenoic acid (DHA); polyunsaturated fatty acid; fatty acid synthesis pathway; polyketide synthase pathway

1. Introduction

Long-chain polyunsaturated fatty acids (LC-PUFAs), such as docosahexaenoic acid (DHA, 22:6n-3), eicosapentaenoic acid (EPA, 20:5n-3), and arachidonic acid (ARA, 20:4n-6), have gained increasing attention because of their potential health benefits to humans. These benefits have been demonstrated in a number of previous studies, and include the proper development of infant brain and eyes [1,2]; cardiovascular disease prevention [3]; antioxidative, anti-inflammatory [4] as well as anti-cancer properties [5], and the prevention of Alzheimer's disease [6]. Currently, ocean fish are the primary PUFA production source. However, because of the drawbacks of a fishy odor, the accumulation of contaminants, and declining fish stock, it is necessary to identify other sustainable and relatively safer alternative sources of PUFA [7].

Thraustochytrids, a single-celled eukaryotic marine protist belonging to the class Labyrinthulomycetes, can accumulate large amounts of DHA, resulting in these organisms having attracted a great deal of scientific and industrial interest. Research has shown that some thraustochytrid strains can be cultivated to yield high biomasses containing substantial quantities of lipids abundant in PUFAs [8]. There is an abundance of research on the optimization of fermentation parameters in terms of salinity [9], pH, temperature [10], and cultivation medium [11] for high DHA production. Besides, metabolic engineering is also used as a promising approach to promote DHA productivity. Recent research has indicated that DHA is synthesized by two distinct pathways in thraustochytrid: the fatty acid synthase (FAS) pathway and the polyketide synthase (PKS) pathway [12]. The standard FAS pathway synthesizes fatty acids through a series of elongase- and desaturase-catalyzed reactions. Delta-4 desaturase, delta-5 desaturase, and delta-12 desaturase have been reported in *Thraustochytrium aureum* ATCC 34304 [13–15], and delta-5 elongase, delta-6 elongase, and delta-9 elongase have also been successfully identified in some thraustochytrid strains [16,17]. Fatty acids are synthesized through the PKS pathway via highly repetitive cycles of four reactions, including condensation by ketoacyl synthase (KS), ketoreduction (KR), dehydration, and enoyl reduction (ER). Meesapyodsuk and Qiu identified three large subunits of a type I PKS-like PUFA synthase in *Thraustochytrium* sp. 26185 [18]. Nevertheless, to date, the exact biosynthetic mechanism of DHA in *Thraustochytrid* species remains unknown.

De novo genome assembly enables functional genomics research in species whose genetic information is limited, especially in many "non-model" organisms. However, for the *Thraustochytrids* species, the available genomic information is limited. Ye et al. reconstructed a genome-scale metabolic map of *Schizochytrium limacinum* SR21 and found that the biosynthesis of DHA via the PKS pathway requires abundant acetyl-CoA and NADPH, and that 30 genes were predicted as potential targets for DHA overproduction [19]. Comparative genomic analysis revealed that both the FAS and PKS pathways of PUFA production were incomplete and that secreted carbohydrate-active enzymes were remarkably consumed in the genomes of two newly isolated *Thraustochytrids* strains, Mn4 and SW8 [20]. Genome sequence and analysis of *Thraustochytrium* sp. 26185 also showed that both aerobic and anaerobic pathways are present in this strain. However, an essential gene encoding stearate delta-9 desaturase was missing in the aerobic pathway.

Thraustochytriidae sp. SZU445 is a promising DHA-producing candidate with DHA production levels up to 45% of the dry cell weight. In this study, the high-quality genome assembly of *Thraustochytriidae* sp. SZU445 is reported using combined second- and third-generation sequencing strategies. The comprehensive genomic analyses provide new insight into PUFA biosynthesis and molecular machinery, laying a framework for future genetic studies and the biotechnological utility of oleaginous microorganisms.

2. Results

2.1. Genome Sequencing and De Novo Assembly

A total of 64.71 Mb raw data were generated by PacBio platform (BGI, Shenzhen, China) and MGI2000 platform (MGI, Shenzhen, China). After shortening the adaptor sequences and filtering out low-quality data, 63.55 Mb of clean data was obtained. The PacBio platform was then used to generate high-quality subreads. From 1,055,214 subreads, 11,245,510,279 bp were obtained with 25 scaffolds and 54 contigs generated after sequence assembly (Table 1). The N50 values of the scaffolds and contigs were 5,997,638 bp and 2,552,134 bp, respectively. The GC content of the scaffolds and contigs was 45.04%. The Poisson distributions of the GC content and GC depth analysis (Figure 1) suggested that there was no external DNA contamination or sequencing bias, demonstrating that the genome assembly was of high quality. In addition, according to the sequence of the sequenced marine fungus, the (G − C)/(G + C) calculation method was used for GC skew analysis, and based on the results of gene distribution, ncRNA distribution, and annotation, a gene circle was constructed that indicated the distribution of each component in the genome (Figure 2). The genome of *Thraustochytriidae* sp. SZU445 (Table 2) contains approximately 14,145 genes, 18,768 exons, 14,145 CDSs, and 4623 introns. The total length of genes is 26,947,341 bp, with an average length of 1905.08 bp. Among all genes, those with lengths of 200–499 nt and 500–999 nt were drastically more abundant than those of other lengths (Figure 3). Based on the assembly and annotation results, 1106 copies of the noncoding RNA were predicted, which included nearly 493 copies (0.06%) of tRNA, 235 copies (0.64%) of rRNA, and 77 copies (0.01%) of snRNA (Table 3).

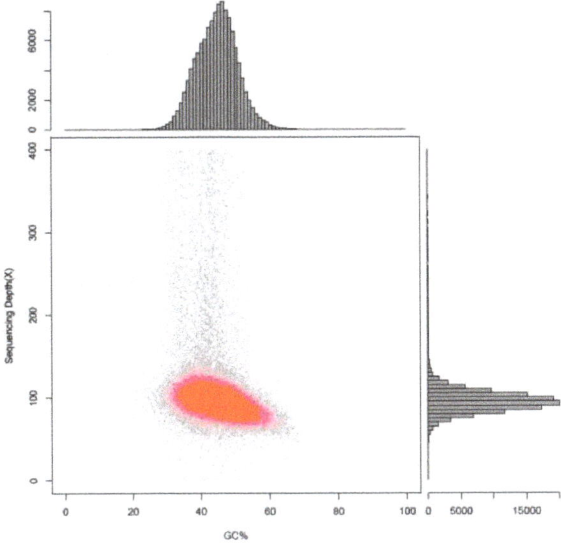

Figure 1. Statistical analysis of the GC content and depth correlation analysis of *Thraustochytriidae* sp. SZU445. The abscissa is the GC content, and the ordinate is the average depth. The scatter plot shows a shape that approximates a Poisson distribution and shows that sequencing data have low GC bias.

2.2. Genome Sequence Annotation

All genes of *Thraustochytriidae* sp. SZU445 were aligned to sequences in 18 public databases, including the Carbohydrate-Active enZYmes Database (CAZY), the Transporter Classification Database (TCDB) database, Swiss-Prot, InterPro, GO, KOG/COG, and KEGG (Table 4). The number and percentage of genes annotated by each database are presented in Table 4. Notably, 30.35% of the genes were unidentifiable, the percentage of which was similar to that observed in other species [21–23].

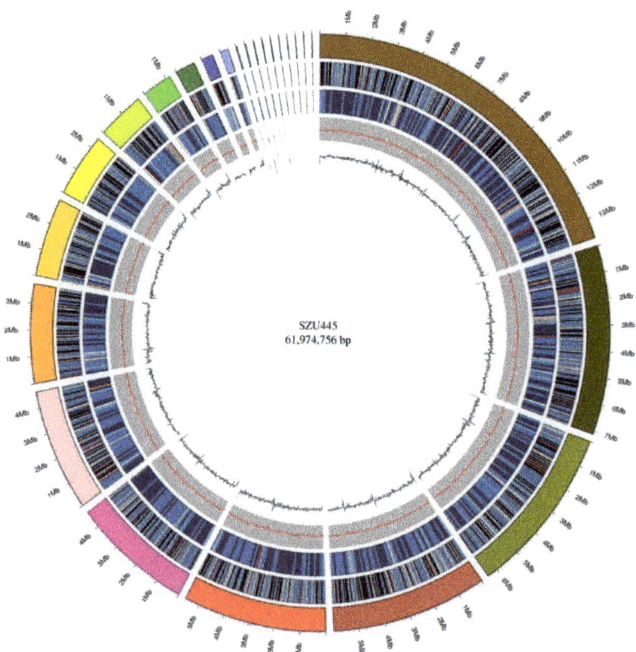

Figure 2. Genomic circle diagram of *Thraustochytriidae* sp. SZU445. From the outer to the inner rings: Genome (sorted by length), gene density (gene number in 50,000 bp nonoverlapping windows), ncRNA density (ncRNA number in 100,000 bp nonoverlapping windows), GC (GC rate in 20,000 bp nonoverlapping windows), GC_skew (GC skew in 20,000 bp nonoverlapping windows).

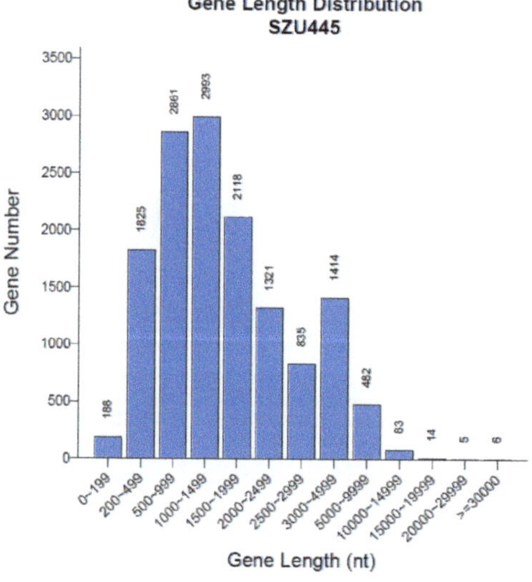

Figure 3. Gene length distribution of *Thraustochytriidae* sp. SZU445. The abscissa is the length of the gene, and the ordinate is the number of genes corresponding to the length of the gene.

Table 1. Summary of clean data assembly for *Thraustochytriidae* sp. SZU445.

Sample Name	Seq Type (#)	Total Number (#)	Total Length (Mb)	N50 Length (Mb)	N90 Length (Mb)	Max Length (Mb)	Min Length (Mb)	Gap Number (Mb)	GC Content (%)
SZU445	Scaffold	25	61.97	5.98	2.41	13.75	0.0054	0.091	45.04
SZU445	Contig	54	61.88	2.55	1.39	4.00	0.0054	-	45.04

Seq Type (#): Sequence type (Scaffold, Contig). Total number (#): Total number of Contig or Scaffold. Total length (Mb): Total length of assembly results. N50 Length (Mb): The N50 length is used to determine the assembly continuity; the higher the better. N50 is a weighted median statistic that 50% of the total length is contained in transcripts that are equal to or larger than this value. N90 length (Mb): similar to N50 length. Max length (Mb): max length of scaffold or contig. Min length (Mb): min length of scaffold or contig. Gap number (Mb): number of gaps in the sequence. GC content (%): the percentage of G and C bases in the assembly result sequence.

Table 2. The Genetic component of *Thraustochytriidae* sp. SZU445.

Sample Name (#)	Type (#)	Total Number (#)	Total Length (bp)	Average Length (bp)	Length/Genome Length (%)
SZU445	Gene Stat	14,145	26,947,341	1905.08	43.48
	Exons Stat	18,768	25,518,500	1359.68	41.18
	CDS Stat	14,145	25,518,500	1804.07	41.18
	Intron Stat	4623	1,428,841	309.07	2.31

Table 3. Noncoding RNA statistics of *Thraustochytriidae* sp. SZU445.

Sample Name (#)	Type	Copy#	Avg_Len	Total_Len	% in Genome
SZU445	tRNA	493	77.81	38,362	0.0619
	rRNA	235	1683.92	395,723	0.6385
	snRNA	77	70.15	5402	0.0087

Type: ncRNA type. Copy: The number of ncRNA type copies. Avg_Len: The average length of ncRNA. Total_Len: The total length of ncRNA. % in Genome: ncRNA types as a percentage of the genome.

Table 4. Gene annotation results of *Thraustochytriidae* sp. SZU445 according to the database.

Total	CAZY	TCDB	IPR	SWISS-PROT	GO	KEGG	KOG	COG	P450	TF	EKPD	NOG	CARD	CWDE	NR	DBCAN	PHI	PHOSPHATASE	Overall
14,145	34 (0.24%)	232 (1.64%)	9707 (68.62%)	2122 (15%)	7255 (51.29%)	1625 (11.48%)	1866 (13.19%)	1324 (9.36%)	740 (5.23%)	364 (2.57%)	369 (2.60%)	3550 (25.09%)	7 (0.04%)	1 (0.7%)	2629 (18.58%)	207 (1.46%)	412 (2.91%)	85 (0.60%)	9852 (69.65%)

CAZY: Carbohydrate-Active enZYmes Database. TCDB: Transporter Classification Database. IPR: InterPro Database. GO: Gene Ontology Database. KEGG: Kyoto Encyclopedia of Genes and Genomes. KOG: EuKaryotic Orthologous Groups. COG: Clusters of Orthologous Groups. P450: Fungal Cytochrome P450 Database. TF: Transcription Factor database. EKPD: Eukaryotic Protein Kinases and Protein Phosphatases. NOG: Evolutionary genealogy of genes: Non-supervised Orthologous Groups. CARD: The Comprehensive Antibiotic Resistance Database. CWDE: Cell Wall Degrading Enzyme. NR: Non-Redundant Protein Database. DBCAN: a web server and Database for automated Carbohydrate-active enzyme ANnotation. PHI: Pathogen–Host Interactions. PHOSPHATASE: a phosphatases database of EKPD.

The GO database was founded by the Gene Ontology Association in 1988 including three categories: (1) cellular component, used to describe locations, subcellular structures, and macromolecular complexes such as nucleoli, telomeres, and the recognition of the initial complex; (2) molecular function, used to describe the gene functions and gene products, such as binding to carbohydrates or ATP hydrolase activity; and (3) biological process, used to describe the ordered combination of molecular functions to achieve broader biological functions, such as mitosis [24]. Genes are assigned to one or more of these categories depending on the nature of the product. Through the GO database annotation, we speculated the possible functions of genes based on their annotations in different categories. In *Thraustochytrium* sp. SZU445, 22,446 genes were annotated based on the GO database (Figure 4). The majority genes in the biological process category were involved in cellular and metabolic processes, with 3084 and 2634 genes identified, respectively. Simultaneously, only one gene participated in cell proliferation and detoxification in the cellular component category. For the cellular component category, cell (1027 genes), cell part (1027 genes), and membrane (872 genes) were the top three terms. In the molecular function category, the number of genes participating in binding (4506 genes) and catalytic activity (3305 genes) occupied the dominant position. Only one gene related to metallochaperone activity was identified in the molecular function category. According to the literature, fatty acid synthesis-related genes are primarily classified in the metabolic process group of biological process [25,26]. Thus, fatty acid anabolic processes dominated the biological processes of *Thraustochytrium* sp. SZU445.

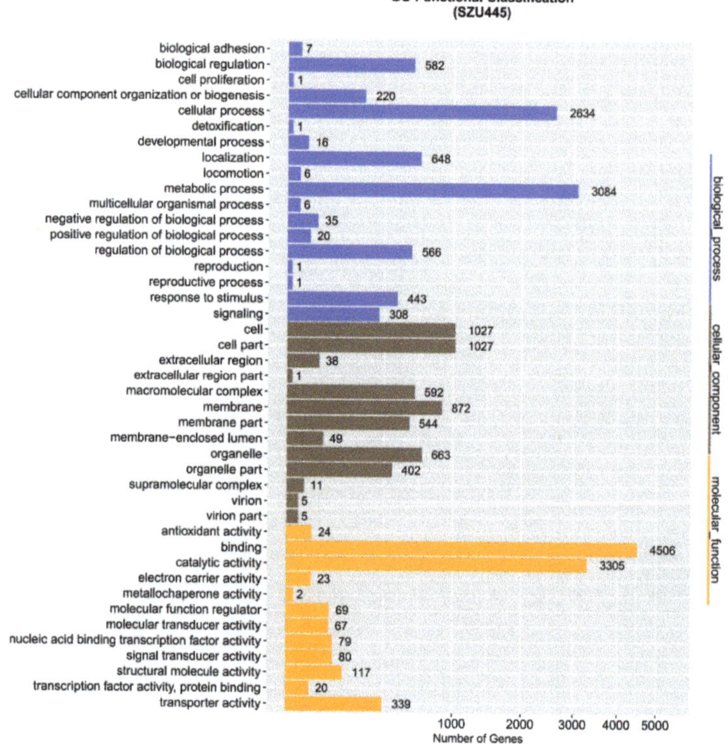

Figure 4. Distribution of GO database functional annotations. The ordinate is the annotation item, and the abscissa is the number of corresponding genes.

KEGG is a database of large-scale molecular data generated from molecular-level information—particularly genome sequencing and other high-throughput experimental techniques—to understand the advanced functions and utility of biological systems. The database divides biological pathways into eight categories, each of which has subdivisions. Each subdivision is annotated with an associated gene that is then graphically displayed. Through this database annotation, it is easy to find genes of all annotations related to a specific function. According to the results, 3513 genes were assigned into 6 KEGG classifications (cellular processes, environmental, genetic, human diseases, metabolism, and organismal systems) in *Thraustochytrium* sp. SZU445 (Figure 5). The percentages of genes involved in the 6 KEGG classifications (metabolism and organismal systems, environmental, human diseases, genetic, cellular processes) were 10.84%, 5.65%, 13.13%, 20.66%, 34.10%, and 15.74%, respectively. Among them, the dominant classification was genes involved in metabolism pathways (1198 genes). The metabolism classification can also be divided into 12 subclassifications (metabolism of amino acid, biosynthesis of other secondary, carbohydrate, energy, lipid, cofactors and vitamins, other amino acids, terpenoids and polyketides, nucleotide, global and overview maps, glycan biosynthesis and metabolism, xenobiotic biodegradation, and metabolism). Sixty-eight genes participated in lipid metabolism, which was fifth among the remaining subclassifications. This result provided the foundation for further gene analysis of fatty acid metabolism. In particular, the number of genes involved in global and overview maps (430 genes) was the largest among all subclassifications and classifications.

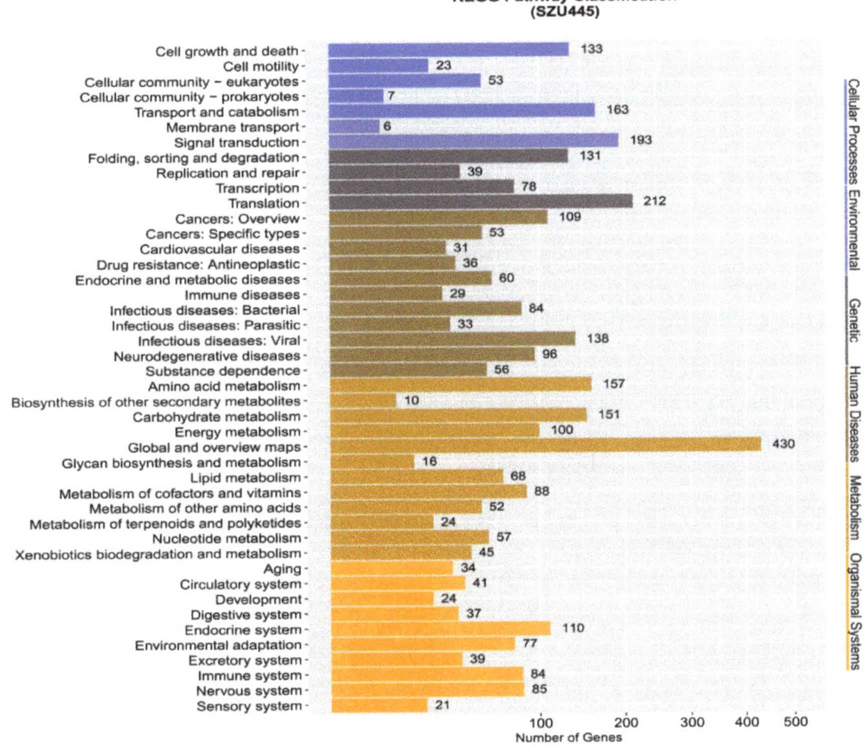

Figure 5. Distribution of KEGG database functional annotations. The ordinate is the annotation item, and the abscissa is the number of corresponding genes.

2.3. Phylogenetic Analysis of Thraustochytriidae sp. SZU445

Based on the 18S rRNA sequences of the sample Thraustochytriidae sp. SZU445 and the reference strains, we used the MEGA software to construct a phylogenetic tree. The neighbor-joining method was applied for the analysis. As shown in Figure 6, Thraustochytriidae sp. SZU445 was shown to share a common ancestor with all reference strains. The evolutionary position of SZU445 is lower compared with the reference strains, while it is higher than Aurantiochytrium sp. HS399. In terms of kinship, Thraustochytriidae sp. SZU445 showed a higher kinship than Aurantiochytrium sp. LY-2012.

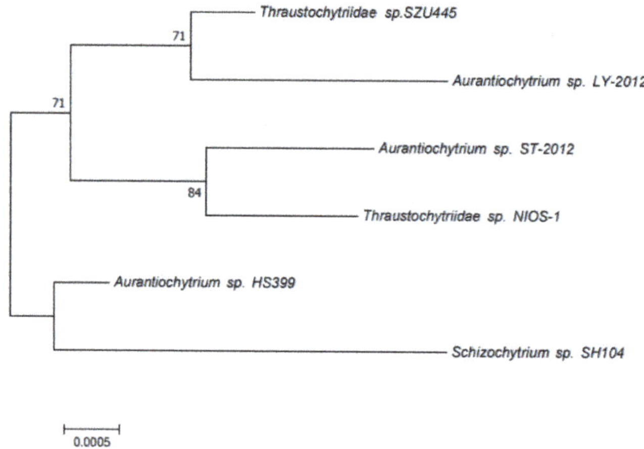

Figure 6. The phylogenetic tree produced using the neighbor-joining method analysis. The evolutionary history was inferred using the neighbor-joining method. The optimal tree with the sum of branch length = 0.01309240 is shown. The percentage of replicate trees in which the associated taxa clustered together in the bootstrap test (1000 replicates) is shown next to the branches. The evolutionary distances were computed using the maximum composite likelihood method and are in the units of the number of base substitutions per site. The analysis involved six nucleotide sequences. Codon positions included were 1st + 2nd + 3rd + noncoding. All ambiguous positions were removed for each sequence pair. There were a total of 1739 positions in the final dataset. Evolutionary analyses were conducted in MEGA7.

2.4. Analysis of Genes Involved in Long-Chain Fatty Acid (LCFA) Biosynthesis

We selected the genes that participated in LCFA biosynthesis from the genome sequencing and protein annotation results in the KEGG database (Table 5). There are two main ways to synthesize LCFAs: The FAS pathway and the PKS pathway.

Table 5. Enzymes involved in fatty acid biosynthesis and metabolism identified by annotation of the *Thraustochytriidae* sp. SZU445 genome.

Enzyme	EC Number	Number of Transcripts
Fatty Acid Desaturation and Elongation		
delta7-sterol 5-desaturase	1.14.19.20	1
sphingolipid 8-(E)-desaturase	1.14.19.18	1
sphingolipid 4-desaturase	1.14.19.17 1.14.18.5	1
aldehyde dehydrogenase (NAD+)	1.2.1.3	4
17beta-estradiol 17-dehydrogenase	1.1.1.62 1.1.1.330	1
acyl-CoA dehydrogenase	1.3.8.7	145

Table 5. *Cont.*

Enzyme	EC Number	Number of Transcripts
Fatty Acid Desaturation and Elongation		
glycerol-3-phosphate dehydrogenase	1.1.5.3	26
S-(hydroxymethyl)glutathione dehydrogenase/alcohol dehydrogenase	1.1.1.284 1.1.1.1	2
glycerol-3-phosphate dehydrogenase	1.1.5.3	26
glutaryl-CoA dehydrogenase	1.3.8.6	5
glycerol-3-phosphate dehydrogenase (NAD+)	1.1.1.8	1
alcohol dehydrogenase (NADP+)	1.1.1.2	9
aldehyde dehydrogenase family 7 member A1	1.2.1.31 1.2.1.8 1.2.1.3	2
3-hydroxyacyl-CoA dehydrogenase	1.1.1.35	29
glycerol 2-dehydrogenase (NADP+)	1.1.1.156	1
S-(hydroxymethyl)glutathione dehydrogenase/alcohol dehydrogenase	1.1.1.284 1.1.1.1	2
17beta-estradiol 17-dehydrogenase/very-long-chain 3-oxoacyl-CoA reductase	1.1.1.62 1.1.1.330	1
delta14-sterol reductase	1.3.1.70	1
Fatty Acid Biosynthesis		
acetyl-CoA acyltransferase 2	2.3.1.16	2
acetyl-CoA acyltransferase 1	2.3.1.16	2
hydroxymethylglutaryl-CoA synthase	2.3.3.10	6
fatty acid synthase subunit alpha	2.3.1.86	2
3-oxoacyl-[acyl-carrier-protein] synthase II	2.3.1.179	1
acetyl-CoA carboxylase/biotin carboxylase 1	6.4.1.2 6.3.4.14 2.1.3.15	1

In *Thraustochytrium* sp. SZU445, FAS (KEGG, EC: 2.3.1.86) (total gene length: 12,443 bp; CDS length: 12,285 bp; number of amino acids: 4147; number of copies: 1) and acyl carrier protein (ACP) (COG, COG ID:COG0236) (total gene length: 429 bp; CDS length: 429 bp; number of amino acids: 143; number of copies: 1) were sequenced and annotated. In the FAS pathway, the acetic acid to long-chain saturated fatty acids (C16:0 and C18:0) step of primary fatty acid biosynthesis is catalyzed by a type I FAS, and long-chain unsaturated fatty acids are generated after a string of desaturation and elongation reactions through desaturases and elongases [27]. FAS is a multifunctional polypeptide of 4147 amino acids that includes four catalytic domains for biochemical reactions: condensation, ketoacyl reduction, hydroxyacyl dehydration, and enoyl reduction [28]. Additionally, it also possesses an ACP domain for binding of the phosphopantetheine prosthetic group to carry an acyl chain. Furthermore, a malonyl-CoA ACP transacylase (MAT) domain, that can catalyze the transformation of malonyl-CoA to malonyl-ACP, was also detected in the enzyme [29]. In the aerobic pathway, the biosynthesis of LCFAs involves a string of desaturation and elongation reactions. Therefore, desaturase and elongase are highly significant in the aerobic FAS pathway. The second half of the FAS pathway can be divided into two subpathways. One subpathway starts with C18:3 (9,12,15), the biosynthesis of which involves the delta-6 desaturation of C18:3 (9,12,15) to C18:4 (6,9,12,15), the elongation of C18:4 (6,9,12,15) to C20:4 (8,11,14,17), and the delta-5 desaturation of C20:4 (8,11,14,17) to C20:5 (5,8,11,14,17) followed by another elongation of C20:5 (5,8,11,14,17) to C22:5 (7,10,13,16,19) and a final delta-4 desaturation of C22:5 (7,10,13,16,19) to C22:6 (4,7,10,13,16,19). The same enzymes are involved in both the first and second pathways. In *Thraustochytrium* sp. SZU445, two desaturases and elongases were detected: delta-4 (KEGG, EC: 1.14.19.17 1.14.18.5) (total gene length: 1471 bp; CDS length: 1167 bp; number of amino acids: 388; number of copies: 1), delta-9 (KEGG, EC: 1.14.19.18) (total gene length: 1601 bp; CDS length: 1601 bp, number of amino acids: 541; number of copies: 1) and elongase (KOG, KOG ID:KOG3072) (total gene length: 1088 bp; CDS length: 1088 bp; number of amino acids: 362; number of copies: 1). These protein products exhibited 100.00%, 95.54%, and 99.87% similarity to the sphingolipid delta-4 desaturase of *Hondaea fermentalgiana* (NCBI accession: GBG25593.1), the delta-9 desaturase of *Hondaea fermentalgiana* (NCBI accession: GBG34675.1), and the elongation of very-long-chain fatty acids protein 6 of *Hondaea fermentalgiana* (NCBI accession: GBG25297.1) [30]. The characterization of this desaturase gene and its function was reported by exogenous expression in yeast [31–33]. However, there were only two desaturases (delta-4 and delta-9 desaturase) annotated in *Thraustochytrium* sp. SZU445 that belonged to the classic FAS pathway. The results indicated that a nonclassic FAS pathway may exist in *Thraustochytrium* sp. SZU445 for LC-PUFA synthesis.

PKS has been identified in prokaryotes and eukaryotes [12]. Briefly, fatty acids and polyketides are constructed by repeated decarboxylative Claisen ester condensations of an acyl-CoA starter part with malonyl-CoA units catalyzed by a 3-ketoacyl-synthase (KS). This process often involves an ACP and a (malonyl) acyltransferase (MAT/AT). In fatty acid biosynthesis, beta-oxidation processing typically yields an entirely saturated acyl backbone by ketoreductase (KR), dehydratase (DH), and an enoyl reductase (ER) [34]. An isomerase changes the conformation of the fatty acid chain. KS (KEGG, EC: 2.3.1.179) (total gene length: 1335 bp; CDS length: 1335 bp; number of amino acids: 444; number of copies: 1), KR (KEGG, EC: 1.1.1.-) (total gene length: 1489 bp; CDS length: 1266 bp; number of amino acids: 422; number of copies: 1), and ER (KEGG, EC: 1.3.1.9) (total gene length: 1146 bp; CDS length: 1146 bp; number of amino acids: 382; number of copies: 1) were detected in *Thraustochytrium* sp. SZU445 by genome sequencing and protein annotation. The KS, KR, and ER were highly homologous to the beta-ketoacyl-acyl-carrier-protein synthase II of *Aphanomyces invadans* (NCBI accession: XP_008875515.1), oxidoreductase of "*Candidatus Poribacteria*" bacterium (NCBI accession: RKU36622.1) [35] and tRNA-dihydrouridine(47) synthase NADP+-like of *Hondaea fermentalgiana* (NCBI accession: GBG27734.1) [30] with approximately 94.45%, 95.67%, and 99.00% identity at the amino acid level. These results demonstrate that *Thraustochytrium* sp. SZU445 may possess the PKS pathway but not the classical pathway. Nevertheless, neither the gene nor protein for the DH and isomerase were discovered in *Thraustochytrium* sp. SZU445. We found two enzymes in *Thraustochytrium* sp. SZU445 with functions that were highly similar to those of a DH and an isomerase. These two enzymes were dTDP-glucose 4,6-DH (KEGG, EC: 4.2.1.46) (total gene length: 1134 bp; CDS length: 1134 bp; number of amino acids: 377; number of copies: 1) and delta-3,5-delta-2,4-dienoyl-CoA isomerase (KEGG, EC: 5.3.3.21) (total gene length: 852 bp; CDS length: 852 bp; number of amino acids: 283; number of copies: 1). Both of these proteins were 94.00% and 99.00% homologous to the dTDP-glucose 4,6-DH of *Fistulifera solaris* (NCBI accession: GAX23278.1) [36] and delta-3,5-delta-2,4-dienoyl-CoA isomerase of *Hondaea fermentalgiana* (NCBI accession: GBG25986.1) [30]. Thus, we propose that these two enzymes are the isozymes of DH and isomerase and that they participate in the PKS pathway together with KS, ER, and ER. The specific functions of dTDP-glucose 4,6-DH and delta-3,5-delta-2,4-dienoyl-CoA isomerase are discussed in the next section.

3. Discussion

3.1. Fatty Acid Synthesis by the FAS Pathway

In the FAS pathway, the synthesis of PUFAs can be divided into two parts [37]. In the first part, palmitic acid (C16:0) is synthesized by the dehydration of acetyl-CoA and malonyl-COA by the FAS complex (NOG, NOG ID: 49.09) [38]. In the second part, palmitic acid (C16:0) is transformed into a very-long-chain (C22) fatty acid by various enzymes. The FAS complex consists of an ACP (COG, COG ID: COG0236) and six enzyme monomers [39]. The six enzyme monomers are acetyl-CoA-ACP transacylase, malonyl-CoA-ACP transacylase, beta-ketoacyl-ACP synthase, beta-ketoacyl-ACP reductase, beta-hydroxyacyl-ACP, DH, enoyl-ACP reductase, and palmitoyl-ACP thioesterase [40]. Then, after a string of desaturation and elongation reactions by delta-9, delta-12, delta-15, delta-6, and delta-4 desaturases [31] and elongases, palmitic acid obtains double bonds, extending the carbon chain and finally generating PUFAs. In *Thraustochytrium* sp. SZU445, FAS (NOG, NOG ID: 49.09) and ACP (COG, COG ID: COG0236) were identified. Therefore, the saturated fatty acids before C18:0 can be synthesized. Interestingly, *Thraustochytrium* sp. SZU445 was not observed to contain delta-12, delta-15, or delta-6 desaturases by whole-genome sequencing and protein annotation; however, it could synthesize PUFAs. A similar phenomenon has also been reported in a number of other studies, where *Arthrospira platensis* was observed to only have 15-*cis*-phytoene desaturase and zeta-carotene desaturase, and *Strongylocentrotus nudus* was observed to only have delta-5, delta-6, and delta-9 desaturases [41–44]. This result may indicate that the FAS pathway is not the primary pathway responsible for PUFA synthesis.

3.2. Fatty Acid Synthesis by the PKS Pathway

Polyketides, the secondary metabolites that harbor multiple units of ketide groups (-CH$_2$-CO-) [45], are synthesized by PKS, which is an enzyme similar to FAS in bacteria [12]. Similar to the FAS pathway, by using ACP as a covalent connection of PUFA synthesis [39,46] (COG, COG ID: COG0236), the PKS pathway proceeds with repeated cycles. The whole cycle includes condensation by using an acyl-ACP and a malonyl-ACP to form a ketoacyl-ACP with KS (KEGG, EC: 2.3.1.179), keto-reduction by converting ketoacyl-ACP to hydroxyacyl-ACP by KR (KEGG, EC: 1.1.1.-), and dehydration by removing a water molecule from hydroxyacyl-ACP. One whole cycle can produce an unsaturated enoyl-ACP, which is then reduced to a saturated acyl chain by ER (KEGG, EC: 1.3.1.9) [47]. However, unlike the FAS pathway, the PKS pathway often neglects steps of the whole cycle, e.g., dehydration and reduction. Therefore, the products of the PKS pathway vary greatly in structure and often contain keto and hydroxyl groups and double bonds. KS, KR, and ER were discovered in *Thraustochytrium* sp. SZU445, but an isomerase and a DH were not identified; however, *Thraustochytrium* sp. SZU445 can also generate PUFAs such as DHA (C22:5) [42–44].

3.3. Putative Fatty Acid Synthesis Pathway in Thraustochytrium sp. SZU445

Since *Thraustochytrium* sp. SZU445 lacked complete FAS and PKS pathways, we speculated that the synthesis of SZU445 LC-PUFAs involves a combination of the two pathways. According to the whole-genome sequencing and protein annotation results, no DH or isomerase genes were found in *Thraustochytrium* sp. SZU445. However, we identified two enzymes with functions that were highly similar to those of DH and isomerase. The two enzymes are dTDP-glucose 4,6-DH (KEGG, EC: 4.2.1.46) and delta-3,5-delta-2,4-dienoyl-CoA isomerase (KEGG, EC: 5.3.3.21). dTDP-glucose 4,6-DH can catalyze the transformation of dTDP-glucose into dTDP-4-keto-6-deoxyglucose, forming a ketone group [48]. The isomerization of 3-*cis*-octenoyl-CoA to 2-*trans*-octenoyl-CoA can be catalyzed by delta-3,5-delta-2,4-dienoyl-CoA isomerase with specific activity for altering the conformation. Moreover, delta-3,5-delta-2,4-dienoyl-CoA isomerase is indispensable for the beta-oxidation of PUFAs [49]. These two enzymes may be the isoenzymes of DH and isomerase and, hence, perform the same functions.

The protein annotation of *Thraustochytrium* sp. SZU445 revealed that the DH and isomerase of the classic PKS pathway are not present in *Thraustochytrium* sp. SZU445. In addition, only the delta-4 and delta-9 desaturases of the classic FAS pathway were found. Moreover, the fatty acid composition of *Thraustochytrium* sp. SZU445 was analyzed using gas chromatography and shown to primarily include methyl tetradecanoate (C14:0), methyl hexadecanoate (C16:0), DPA, and DHA C20:4 (7,10,13,16). Based on these results, we speculated upon the possible synthesis pathway of long-chain fatty acids in *Thraustochytrium* sp. SZU445.

We inferred that long-chain saturated fatty acids of *Thraustochytrium* sp. SZU445, such as methyl hexadecanoate, are produced by the FAS pathway and that long-chain unsaturated fatty acids, such as DHA, are produced by the PKS pathway, with one desaturase participating in this process instead of the desaturation and elongation steps in the FAS pathway (Figure 6). For the long-chain saturated fatty acids, acetyl-CoA and malonyl-CoA are used as substrates to yield methyl tetradecanoate (C14:0) and methyl hexadecanoate (C16:0) by cycling through the FAS pathway six and seven times, respectively. For the long-chain unsaturated fatty acid DPA, C22:4 (7,10,13,16) is generated by cycling through the PKS pathway ten times. C22:5 (4,7,10,13,16) is then converted from C22:4 (7,10,13,16) via delta-4 desaturase. DPA is eventually produced by the activity of delta-3,5-delta-2,4-dienoyl-CoA isomerase (KEGG, EC: 5.3.3.21). In the biosynthesis of DHA, C20:4 (7,10,13,16) is produced by circulating through the PKS pathway nine times. C20:5 (4,7,10,13,16) is formed through the addition of a double bond on the carbon at the fourth position by delta-4 desaturase. Finally, DHA biosynthesis is completed after one additional cycle of the PKS pathway (Figure 7).

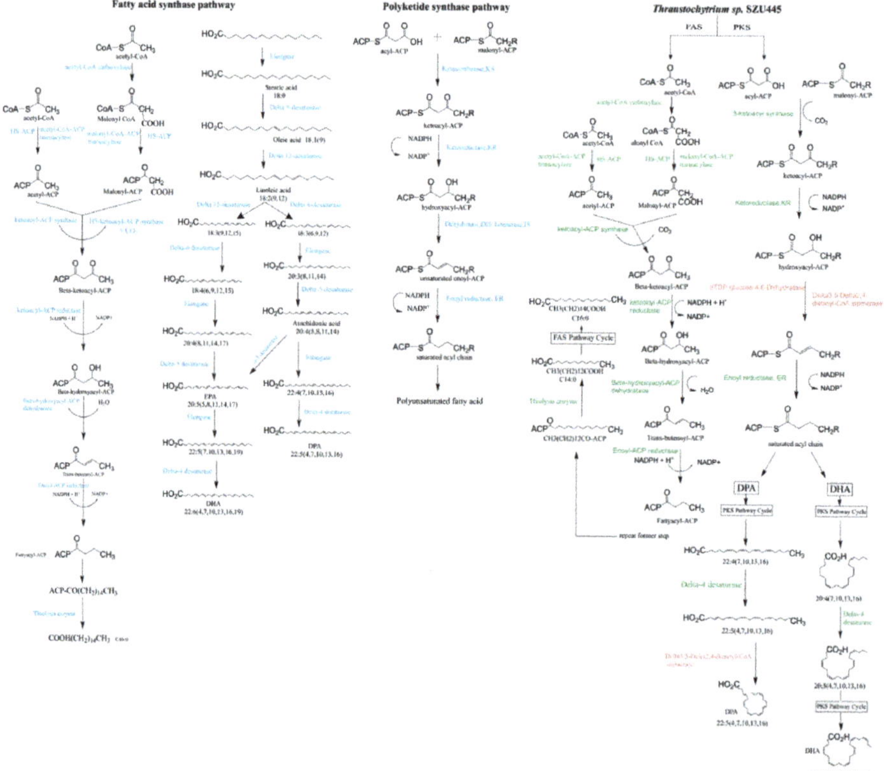

Figure 7. Putative fatty acid synthase pathway of *Thraustochytriidae* sp. SZU445 with the classical fatty acid synthesis pathway and the polyketide synthase pathway. The enzymes colored blue are present in the classical fatty acid synthase (FAS) and polyketide synthase (PKS) pathways. The enzymes colored green are present in *Thraustochytriidae* sp. SZU445 and correspond to the classical FAS and PKS pathways. The enzymes colored red are isozymes of dehydrase and isomerase in the PKS pathway that exist in *Thraustochytriidae* sp. SZU445.

4. Materials and Methods

4.1. Microbes and Cultivation

Thraustochytrium sp. SZU445 was isolated from mangroves (22°31′13.044″ N, 113°56′58.560″ E) in the coastal waters of southern China. Because of the results of an initial assessment for DHA production, *Thraustochytrium* sp. SZU445 was selected for further research. Furthermore, *Thraustochytrium* sp. SZU445 was stored in the China General Microbiological Culture Center (CGMCC) under the accession no. 8095. First, the strain was inoculated into M4 liquid medium, which was prepared in 100% filtered natural seawater containing 0.15% peptone, 2% glucose, 0.025% KH_2PO_4, and 0.1% yeast extract. Seawater was gathered from Mirs Bay, Shenzhen (22°31′32.632″ N, 114°28′40.185″ E), with a salinity of approximately 30.75–33.19‰. Cultures were incubated for 48 h at a constant temperature of 27 °C and with shaking at 180 rpm. Subsequently, 4% of the culture was used to inoculate fresh M4 medium and incubated for three days under the same growth conditions.

4.2. DNA Preparation and Sequencing

The sequencing platform chosen for this subject was a combination of the PacBio (BGI, Shenzhen, China) and MGI2000 (MGI, Shenzhen, China) platforms. Genomic DNA was extracted from a 50 mL fresh cell suspension of *Thraustochytrium* sp. SZU445, which was centrifuged at 8000 rpm for 10 min. The resulting cell pellets were ground to a fine powder in liquid nitrogen, then washed once in 5.0 mL DNA extraction buffer. Subsequently, the pellets were resuspended in 5.0 mL phenol/chloroform/isoamyl alcohol (25:24:1) and centrifuged twice at 9000 rpm for 16 min. The supernatant was then washed in an equal volume of chloroform and centrifuged at 12,000 rpm for 16 min. Genomic DNA was precipitated by adding 2.5 volumes of 100% ethanol and collected by centrifugation at 12,000 rpm for 15 min at 4 °C. The genomic DNA of *Thraustochytrium* sp. SZU445 was sequenced on the PacBio sequencing platform based on third-generation sequencing technologies. Prior to library construction, the concentration, integrity, and purity of the genomic DNA was tested. The concentration was detected using a Qubit fluorometer (Invitrogen, Carlsbad, CA, USA). The purity and integrity of the sample was assessed by agarose gel electrophoresis (agarose gel concentration: 1.2%; voltage: 120 V; electrophoresis time: 50 min). To construct the library for the PacBio platform, 10 µg of genomic DNA was randomly fragmented by g-TUBE (Covaris, Woburn, MA, USA). The fragmented genomic DNA was then processed using an Agencourt AMPure XP-Medium kit (Beckman Coulter, Brea, CA, USA) to obtain fragments with a size of 10 kb. The DNA fragments generated from exonuclease ExoVII (New England Biolabs, Ipswich, MA, USA) digestion and subject to end repair were connected to the hairpin structure at both ends to form a dumbbell structure called SMRTbell. The library for the PacBio sequencing platform was successfully constructed after the purification and sorting of fragments. Qualified library fragments were sequenced after the annealing and binding polymerase steps. The genome sequencing data of *Thraustochytriidae* sp. SZU445 were uploaded to the National Center for Biotechnology Information Search database (NCBI) (BioProject ID: PRJNA579065; BioSample accession: SAMN13135800; SRA accession: PRJNA579065). The method for constructing the library using the MGI2000 platform (MGI, Shenzhen, China) is essentially the same as that described for the PacBio sequencing platform (BGI, Shenzhen, China).

4.3. Fragment Assembly and Gene Annotation

Since the raw sequencing data contained low-quality sequences, such as adaptors as well as duplicated and low-quality reads, these reads were filtered to obtain reliable subreads for assembly and guarantee the reliability of the subsequent analysis results. Genome assembly can be divided into the following four parts. (1) Subreads correction: Use Proovread (Version: 2.12; Parameter setting: -t 4 –coverage 60 –mode sr; related website: https://github.com/BioInf-Wuerzburg/proovread) to perform mixed correction on Subread to get highly reliable corrected reads; (2) corrected reads assembly: based on the corrected reads, Celera (Version: 8.3; parameter setting: doTrim_initialQualityBased = 1, doTrim_finalEvidenceBased = 1, doRemoveSpurReads = 1, doRemoveChimericReads = 1, - d properties -U; related website: http://sourceforge.net/projects/wgs-assembler/files/wgs-assembler/wgs-8.3/) and Falcon were each used to generate separate assemblies. Finally, the optimal assembly result was selected; (3) assembly result correction: single-base correction by GATK (Version: v1.6-13; parameter settings: -cluster 2 -window 5 -stand_call_conf 50 -stand_emit_conf 10.0 -dcov 200 MQ0 ≥ 4; related website: http://www.broadinstitute.org/gatk/) for assembly using second-generation small fragment data to obtain highly reliable assembly sequences; (4) scaffold and hole filling: a scaffold was constructed using SSPACE_Basic_v2.0 (Version: v2.0; related website: http://www.baseclear.com/genomics/bioinformatics/basetools/SSPACE) on the assembly results based on the second-generation Illumina large-segment data, and pbjelly2 (Version: 15.8.24; related website: https://sourceforge.net/projects/pb-jelly/) was used to fill holes in the scaffold. After completing the genome assembly, we first constructed the gene model for subsequent gene function annotation. We chose the de novo prediction to build the gene model. The de novo prediction uses the GeneMark-ES software (Version: 4.21, parameter settings: –genemarkes, Related website: http://exon.gatech.edu/). This software is based on Hidden Markov Model and ab initio algorithm to make predictions, and it does not require reference species and reference sequences. It is self-trained. In the assembled sequence, all possible open reading frames (ORF) were predicted as CDSs by the GeneMarkes software. After the prediction was finished, the predicted ORF was translated into the amino acid sequence. Then it was compared with 18 protein databases such as Kyoto Encyclopedia of Genes and Genomes (KEGG) database (Version: 81), Gene Ontology (GO) (Version: releases_2017-09-08), and Swiss-Prot database (Version: release-2017-07) by BLAST to obtain corresponding functional annotation information. Since each sequence alignment result exceeds one, in order to ensure its biological significance, we retain an optimal alignment result as the annotation of the gene. The Eukaryotic Cluster of Orthologous Groups (KOG) and Kyoto Encyclopedia of Genes and Genomes (KEGG) database pathway annotations were implemented using a separate BLAST search against both the KOG (http://www.ncbi.nlm.nih.gov/COG/) and KEGG (https://www.genome.jp/kegg/) databases. The Gene Ontology (GO) and Swiss-Prot annotations were also carried out using the Swiss-Prot database (https://web.expasy.org/docs/swiss-prot_guideline.html) and the GO database (https://www.ebi.ac.uk/ols/ontologies/go), respectively. In addition, the prediction and annotation of kinase domains in the Eukaryotic Protein Kinases and Protein Phosphatases Database (EKPD) were performed by HMMER (v.3.0) using the default parameters and the kinase database (http://ekpd.biocuckoo.org/faq.php). Moreover, for non-coding RNA prediction, the rRNA was predicted by RNAmmer software (Version: 1.2, parameter settings: –s Species –m Type –gff*.rRNA.gff –f*.rRNA.fq, Related website: http://www.cbs.dtu.dk/services/RNAmmer/). The tRNA region and tRNA secondary structure predicted by tRNAscan-SE software (Version: 1.3.1, parameter settings: –Spec_tag(BAOG) –o *. tRNA –f*.tRNA.structure, Related website: http://gtrnadb.ucsc.edu/). SnRNAs was predicted in comparison with Rfam database (Version: 9.1, parameter settings: –p blastn –W 7 –e 1 –v 10000 –b 10000 –m 8 –i subfile –o *.blast.m8, Related website: http://rfam.sanger.ac.uk/) by using the "covariance model" (CMS) of the Infranal software (Version: 1.1.1 (July 2014), parameter settings: default parameter, Related website: http://rfam.sanger.ac.uk/). The default parameters of Infranal software is displayed in Supplementary Materials.

4.4. Phylogenetic Analysis

The 18S rRNA was selected as the alignment sequence, and the 18S rRNA sequences of the reference strains were obtained from National Center for Biotechnology Information (NCBI). The blast function of the NCBI was used to compare the similarity of *Thraustochytriidae* sp. SZU445 and those of the reference strains. The NCBI accession numbers of the reference strains are shown in Table 6.

Table 6. The National Center for Biotechnology Information (NCBI) accession numbers of the reference strains for the phylogenetic analysis.

Reference Strains	NCBI Accession
Aurantiochytrium sp. HS399	MH319338.1
Aurantiochytrium sp. ST-2012	JQ982490.1
Aurantiochytrium sp. LY-2012	JX847377.1
Schizochytrium sp. SH104	KX379459.1
Thraustochytriidae sp. NIOS-1	AY705769.1

The MEGA software (Version: 7.0.26; algorithm: neighbor-joining method, the maximum composite likelihood method for the evolutionary distances; related website: https://www.megasoftware.net/) was used to construct a phylogenetic tree for *Thraustochytriidae* sp. SZU445 and the reference strains.

5. Conclusions

In summary, this research revealed the essential genome features of *Thraustochytriidae* sp. SZU445 and proposed a putative fatty acid synthesis pathway (specifically for DHA and DPA). Our purpose was to identify metabolism-related genes involved in PUFA synthesis in *Thraustochytriidae* sp. SZU445. This study is the first to provide genome information relating to *Thraustochytriidae* sp. SZU445 using a third-generation sequencing platform. By assembling contigs and scaffolds, we predicted and analyzed the GO and KEGG terms for potential genes and proteins involved in the synthesis of polyunsaturated fatty acids. These data could yield novel insights into the molecular mechanisms of *Thraustochytriidae* sp. and serve in developing innovative strategies for promoting PUFA (including DHA) production.

Supplementary Materials: The default parameters of Infranal software are available online at http://www.mdpi.com/1660-3397/18/2/118/s1.

Author Contributions: X.Z., Y.L., and S.L. (Siting Li) wrote the paper; X.Z., S.L. (Shuangfei Li) designed the concept of the experiments and analyzed the data; L.L. completed the work of strain mutation and provided the strain; C.L., B.W., C.H.K.C., and H.C. checked and revised the detail of the manuscript; X.Y., as the corresponding author, is responsible for critical reading and finalization of the manuscript. All authors have read and agreed to the published version of the manuscript.

Funding: This research was funded by [National Key Research and Development Project] grant number [2018YFA0902500], [the Natural Science Foundation of Guangdong Province] grant number [2018A030313139], [Joint R&D Project of Shenzhen-Hong Kong Innovation] grant number [SGLH20180622152010394], [Natural Science Foundation of Shenzhen] grant number [KQJSCX20180328093806045], [Shenzhen Grant Plan for Science & Technology] grant number [CKCY20160427102211071] and [Shenzhen University Start-up Research Funding] grant number [827-000192, 2016100].

Conflicts of Interest: The authors declare no conflicts of interest.

References

1. Salem, N.; Litman, B.; Kim, H.-Y.; Gawrisch, K. Mechanisms of action of docosahexaenoic acid in the nervous system. *Lipids* **2001**, *36*, 945–959. [CrossRef] [PubMed]

2. Lauritzen, L. The essentiality of long chain n-3 fatty acids in relation to development and function of the brain and retina. *Prog. Lipid Res.* **2001**, *40*, 1–94. [CrossRef]

3. Von Schacky, C.; Harris, W.S. Cardiovascular benefits of omega-3 fatty acids. *Cardiovasc. Res.* **2007**, *73*, 310–315. [CrossRef] [PubMed]

4. Zhang, Y.P.; Brown, R.E.; Zhang, P.C.; Zhao, Y.T.; Ju, X.H.; Song, C. DHA, EPA and their combination at various ratios differently modulated a beta(25-35)-induced neurotoxicity in sh-sy5y cells. *Prostaglandins Leukot. Essent. Fatty Acids* **2018**, *136*, 85–94. [CrossRef]

5. Gillet, L.; Roger, S.; Bougnoux, P.; Le Guennec, J.-Y.; Besson, P. Beneficial effects of omega-3 long-chain fatty acids in breast cancer and cardiovascular diseases: Voltage-gated sodium channels as a common feature? *Biochimie* **2011**, *93*, 4–6. [CrossRef]

6.   Hashimoto, M.; Hossain, S.; Al Mamun, A.; Matsuzaki, K.; Arai, H. Docosahexaenoic acid: One molecule diverse functions. *Crit. Rev. Biotechnol.* **2017**, *37*, 579–597. [CrossRef]

7.   Rubio-Rodríguez, N.; Beltrán, S.; Jaime, I.; De Diego, S.M.; Sanz, M.T.; Carballido, J.R. Production of omega-3 polyunsaturated fatty acid concentrates: A review. *Innov. Food Sci. Emerg. Technol.* **2010**, *11*, 1–12. [CrossRef]

8.   Lewis, T.E.; Nichols, P.D.; McMeekin, T.A. The Biotechnological Potential of Thraustochytrids. *Surg. Endosc.* **1999**, *1*, 580–587. [CrossRef]

9.   She, Z.L.; Wu, L.; Wang, Q.; Gao, M.C.; Jin, C.J.; Zhao, Y.G.; Zhao, L.T.; Guo, L. Salinity effect on simultaneous nitrification and denitrification, microbial characteristics in a hybrid sequencing batch biofilm reactor. *Bioproc. Biosyst. Eng.* **2018**, *41*, 65–75. [CrossRef]

10.   Ma, Z.; Tan, Y.; Cui, G.; Feng, Y.; Cui, Q.; Song, X. Transcriptome and gene expression analysis of DHA producer Aurantiochytrium under low temperature conditions. *Sci. Rep.* **2015**, *5*, 14446. [CrossRef]

11.   Patil, K.P.; Gogate, P.R. Improved synthesis of docosahexaenoic acid (DHA) using Schizochytrium limacinum SR21 and sustainable media. *Chem. Eng. J.* **2015**, *268*, 187–196. [CrossRef]

12.   Metz, J.G.; Roessler, P.; Facciotti, D.; Levering, C.; Dittrich, F.; Lassner, M.; Valentine, R.; Lardizabal, K.; Domergue, F.; Yamada, A.; et al. Production of Polyunsaturated Fatty Acids by Polyketide Synthases in Both Prokaryotes and Eukaryotes. *Science* **2001**, *293*, 290–293. [CrossRef] [PubMed]

13.   Kang, D.H.; Anbu, P.; Kim, W.H.; Hur, B.K. Coexpression of elo-like enzyme and delta 5, delta 4-desaturases derived from *thraustochytrium aureum* atcc 34304 and the production of dha and dpa in pichia pastoris. *Biotechnol. Bioproc. E* **2008**, *13*, 483–490. [CrossRef]

14.   Matsuda, T. Analysis of delta 12-fatty acid desaturase function revealed that two distinct pathways are active for the synthesis of pufas in t Aureum atcc 34304. *J. Lipid Res.* **2012**, *53*, 1210–1222.

15.   Nagano, N.; Sakaguchi, K.; Taoka, Y.; Okita, Y.; Honda, D.; Ito, M.; Hayashi, M. Detection of genes involved in fatty acid elongation and desaturation in thraustochytrid marine eukaryotes. *J. Oleo Sci.* **2011**, *60*, 475–481. [CrossRef]

16.   Hashimoto, K.; Yoshizawa, A.C.; Okuda, S.; Kuma, K.; Goto, S.; Kanehisa, M. The repertoire of desaturases and elongases reveals fatty acid variations in 56 eukaryotic genomes. *J. Lipid Res.* **2008**, *49*, 183–191. [CrossRef]

17.   Ohara, J.; Sakaguchi, K.; Okita, Y.; Okino, N.; Ito, M. Two fatty acid elongases possessing c18-delta 6/c18-delta 9/c20-delta 5 or c16-delta 9 elongase activity in *thraustochytrium* sp. atcc 26185. *Mar. Biotechnol.* **2013**, *15*, 476–486. [CrossRef]

18.   Meesapyodsuk, D.; Qiu, X. Biosynthetic mechanism of very long chain polyunsaturated fatty acids in *Thraustochytrium* sp. 26185. *J. Lipid Res.* **2016**, *57*, 1854–1864. [CrossRef]

19.   Ye, C.; Qiao, W.; Yu, X.-B.; Ji, X.-J.; Huang, H.; Collier, J.; Liu, L. Reconstruction and analysis of the genome-scale metabolic model of schizochytrium limacinum SR21 for docosahexaenoic acid production. *BMC Genom.* **2015**, *16*, 799. [CrossRef]

20.   Song, Z.; Stajich, J.E.; Xie, Y.; Liu, X.; He, Y.; Chen, J.; Hicks, G.R.; Wang, G. Comparative analysis reveals unexpected genome features of newly isolated Thraustochytrids strains: On ecological function and PUFAs biosynthesis. *BMC Genom.* **2018**, *19*, 541. [CrossRef]

21.   Rismani-Yazdi, H.; Haznedaroglu, B.Z.; Bibby, K.; Peccia, J. Transcriptome sequencing and annotation of the microalgae Dunaliella tertiolecta: Pathway description and gene discovery for production of next-generation biofuels. *BMC Genom.* **2011**, *12*, 148. [CrossRef] [PubMed]

22.   Zhao, X.M.; Dauenpen, M.; Qu, C.M.; Qiu, X. Genomic analysis of genes involved in the biosynthesis of very long chain polyunsaturated fatty acids in *thraustochytrium* sp. 26185. *Lipids* **2016**, *51*, 1065–1075. [CrossRef] [PubMed]

23.   Kourmpetis, Y.A.; van der Burgt, A.; Bink, M.C.; ter Braak, C.J.; van Ham, R.C. The use of multiple hierarchically independent gene ontology terms in gene function prediction and genome annotation. *Silico Boil.* **2007**, *7*, 575–582.

24.   Marek, O.; Serge, E.; Antonio, D.S.; Joaquín, D. Evolutionary conservation and network structure characterize genes of phenotypic relevance for mitosis in human. *Plos ONE* **2012**, *7*, e36488.

25.   Speer, N.; Spieth, C.; Zell, A. Spectral Clustering Gene Ontology Terms to Group Genes by Function. *Comput. Vision* **2005**, *3692*, 1–12.

26. Yao, D.; Zhang, C.; Zhao, X.; Liu, C.; Wang, C.; Zhang, Z.; Zhang, C.; Wei, Q.; Wang, Q.; Yan, H.; et al. Transcriptome analysis reveals salt-stress-regulated biological processes and key pathways in roots of cotton (*Gossypium hirsutum* L.). *Genomics* **2011**, *98*, 47–55. [CrossRef]

27. Zuo, G.; Chen, Z.; Jiang, Y.; Zhu, Z.; Ding, C.; Zhang, Z.; Chen, Y.; Zhou, C.; Li, Q. Structural insights into repression of the Pneumococcal fatty acid synthesis pathway by repressor FabT and co-repressor acyl-ACP. *FEBS Lett.* **2019**, *593*, 2730–2741. [CrossRef]

28. Lee, J.Y.; Ha, J.; Yi, J.; Jang, J.; Lee, W.; Lee, Y.; Oh, D.-Y.; Han, K. Superior single nucleotide polymorphisms that contribute to two main routes of the fatty acid synthesis pathway in Korean cattle. *Genes Genom.* **2018**, *40*, 945–954. [CrossRef]

29. Arthur, C.; Williams, C.; Pottage, K.; Płoskoń, E.; PFindlow, S.C.; Burston, S.G.; Simpson, T.J.; Crump, M.P.; Crosby, J. Structure and Malonyl CoA-ACP Transacylase Binding of *Streptomyces coelicolor* Fatty Acid Synthase Acyl Carrier Protein.E. *J. ACS Chem. Boil.* **2009**, *4*, 625–636.

30. Seddiki, K.; Godart, F.O.; Aiese Cigliano, R.; Sanseverino, W.; Barakat, M.; Ortet, P.; Rébeillé, F.; Maréchal, E.; Cagnac, O.; Amato, A. Sequencing, *de novo* assembly, and annotation of the complete genome of a new thraustochytrid species, strain ccap_4062/3. *Genome Announc.* **2018**, *6*, e01335-17. [CrossRef]

31. Qiu, X.; Hong, H.; MacKenzie, S.L. Identification of a delta 4 fatty acid desaturase from *thraustochytrium* sp. Involved in the biosynthesis of docosahexanoic acid by heterologous expression in *Saccharomyces cerevisiae* and *Brassica juncea*. *J. Biol. Chem.* **2001**, *276*, 31561–31566. [CrossRef] [PubMed]

32. Tsai, Y.-Y.; Ohashi, T.; Wu, C.-C.; Bataa, D.; Misaki, R.; Limtong, S.; Fujiyama, K. Delta-9 fatty acid desaturase overexpression enhanced lipid production and oleic acid content in *Rhodosporidium toruloides* for preferable yeast lipid production. *J. Biosci. Bioeng.* **2019**, *127*, 430–440. [CrossRef] [PubMed]

33. Guo, M.; Chen, G.; Chen, J.; Zheng, M. Identification of a Long-Chain Fatty Acid Elongase from *Nannochloropsis* sp. Involved in the Biosynthesis of Fatty Acids by Heterologous Expression in *Saccharomyces cerevisiae*. *J. Ocean Univ. China* **2019**, *18*, 1199–1206. [CrossRef]

34. Kaulmann, U.; Hertweck, C.J.A.C.I.E. Cheminform abstract: Biosynthesis of polyunsaturated fatty acids by polyketide synthases. *Angew. Chem. Int. Ed.* **2002**, *41*, 1866–1869. [CrossRef]

35. Podell, S.; Blanton, J.M.; Neu, A.; Agarwal, V.; Biggs, J.S.; Moore, B.S.; Allen, E.E. Pangenomic comparison of globally distributed poribacteria associated with sponge hosts and marine particles. *ISME J.* **2019**, *13*, 468–481. [CrossRef]

36. Tanaka, T.; Maeda, Y.; Veluchamy, A.; Tanaka, M.; Abida, H.; Maréchal, E.; Bowler, C.; Muto, M.; Sunaga, Y.; Tanaka, M.; et al. Oil Accumulation by the Oleaginous Diatom Fistulifera solaris as Revealed by the Genome and Transcriptome. *Plant Cell* **2015**, *27*, 162–176. [CrossRef]

37. Schonauer, M.S.; Kastaniotis, A.J.; Hiltunen, J.K.; Dieckmann, C.L. Intersection of RNA Processing and the Type II Fatty Acid Synthesis Pathway in Yeast Mitochondria. *Mol. Cell. Boil.* **2008**, *28*, 6646–6657. [CrossRef]

38. Horrobin, D. Nutritional and medical importance of gamma-linolenic acid. *Prog. Lipid Res.* **1992**, *31*, 163–194. [CrossRef]

39. Prescott, D.J.; Vagelos, P.R. Acyl carrier protein. *Adv. Enzymol. Relat. Areas Mol. Biol.* **1972**, *36*, 269–311.

40. Semenkovich, C.F. Regulation of fatty acid synthase (fas). *Prog. Lipid Res.* **1997**, *36*, 43–53. [CrossRef]

41. Kumaresan, V.; Sannasimuthu, A.; Arasu, M.V.; Al-Dhabi, N.A.; Arockiaraj, J. Molecular insight into the metabolic activities of a protein-rich micro alga, *Arthrospira platensis* by de novo transcriptome analysis. *Mol. Boil. Rep.* **2018**, *45*, 829–838. [CrossRef]

42. Hoang, M.H.; Nguyen, C.; Pham, H.Q.; Duc, L.H.; Van Son, L.; Hai, T.N.; Ha, C.H.; Nhan, L.D.; Anh, H.T.L.; Thom, L.T.; et al. Transcriptome sequencing and comparative analysis of Schizochytrium mangrovei PQ6 at different cultivation times. *Biotechnol. Lett.* **2016**, *38*, 1781–1789. [CrossRef]

43. Shah, S.; Sahoo, D.; Shukla, R.N.; Mishra, G. De novo transcriptome sequencing of monodopsis subterranea ccala 830 and identification of genes involved in the biosynthesis of eicosapentanoic acid and triacylglycerol. *Vegetos* **2019**, *32*, 600–608.

44. Wei, Z.; Liu, X.; Zhou, Z.; Xu, J. De novo transcriptomic analysis of gonad of Strongylocentrotus nudus and gene discovery for biosynthesis of polyunsaturated fatty acids. *Genes Genom.* **2019**, *41*, 583–597. [CrossRef] [PubMed]

45. Bentley, R.; Bennett, J.W. Constructing Polyketides: From Collie to Combinatorial Biosynthesis. *Annu. Rev. Microbiol.* **1999**, *53*, 411–446. [CrossRef] [PubMed]

46. Von Wettstein-Knowles, P.; Olsen, J.; McGuire, K.A.; Henriksen, A. Fatty acid synthesis. Role of active site histidines and lysine in Cys-His-His-type beta-ketoacyl-acyl carrier protein synthases. *FEBS J.* **2006**, *273*, 695–710. [CrossRef] [PubMed]

47. Qiu, X. Biosynthesis of docosahexaenoic acid (DHA, 22:6-4, 7,10,13,16,19): Two distinct pathways. *Prostaglandins Leukot. Essent. Fat. Acids* **2003**, *68*, 181–186. [CrossRef]

48. Gross, J.W.; Hegeman, A.D.; Vestling, M.M.; Frey, P.A. Characterization of Enzymatic Processes by Rapid Mix–Quench Mass Spectrometry: The Case of dTDP-glucose 4,6-Dehydratase. *Biochemistry* **2000**, *39*, 13633–13640. [CrossRef]

49. Geisbrecht, B.V.; Zhu, D.; Schulz, K.; Nau, K.; Morrell, J.C.; Geraghty, M.; Schulz, H.; Erdmann, R.; Gould, S.J. Molecular characterization of *Saccharomyces cerevisiae* delta3, delta2-enoyl-coa isomerase. *J. Biol. Chem.* **1998**, *273*, 33184–33191. [CrossRef]

 *marine drugs*

Article

Monogalactosyldiacylglycerol and Sulfolipid Synthesis in Microalgae

Gennaro Riccio [1], Daniele De Luca [2] and Chiara Lauritano [1,*

[1] Department of Marine Biotechnology, Stazione Zoologica Anton Dohrn, CAP80121 Naples, Italy; gennaro.riccio@szn.it

[2] Department of Humanities, Università degli Studi Suor Orsola Benincasa, CAP80135 Naples, Italy; daniele.deluca088@gmail.com

* Correspondence: chiara.lauritano@szn.it; Tel.: +39-081-5833-221

Received: 12 April 2020; Accepted: 27 April 2020; Published: 1 May 2020

Abstract: Microalgae, due to their huge taxonomic and metabolic diversity, have been shown to be a valuable and eco-friendly source of bioactive natural products. The increasing number of genomic and transcriptomic data will give a great boost for the study of metabolic pathways involved in the synthesis of bioactive compounds. In this study, we analyzed the presence of the enzymes involved in the synthesis of monogalactosyldiacylglycerols (MGDGs) and sulfoquinovosyldiacylglycerols (SQDG). Both compounds have important biological properties. MGDGs present both anti-inflammatory and anti-cancer activities while SQDGs present immunostimulatory activities and inhibit the enzyme glutaminyl cyclase, which is involved in Alzheimer's disease. The Ocean Global Atlas (OGA) database and the Marine Microbial Eukaryotic Transcriptome Sequencing Project (MMETSP) were used to search MGDG synthase (MGD), UDP-sulfoquinovose synthase (SQD1), and sulfoquinovosyltransferase (SQD2) sequences along microalgal taxa. *In silico* 3D prediction analyses for the three enzymes were performed by Phyre2 server, while binding site predictions were performed by the COACH server. The analyzed enzymes are distributed across different taxa, which confirms the importance for microalgae of these two pathways for thylakoid physiology. MGD genes have been found across almost all analyzed taxa and can be separated in two different groups, similarly to terrestrial plant MGD. SQD1 and SQD2 genes are widely distributed along the analyzed taxa in a similar way to MGD genes with some exceptions. For Pinguiophyceae, Raphidophyceae, and Synurophyceae, only sequences coding for MGDG were found. On the contrary, sequences assigned to Ciliophora and Eustigmatophyceae were exclusively corresponding to SQD1 and SQD2. This study reports, for the first time, the presence/absence of these enzymes in available microalgal transcriptomes, which gives new insights on microalgal physiology and possible biotechnological applications for the production of bioactive lipids.

Keywords: microalgae; monogalactosyldiacylglycerol synthase; UDP-sulfoquinovose synthase; sulfoquinovosyltransferase; monogalactosyldiacylglycerols; sulfoquinovosyldiacylglycerols; transcriptome analysis

---

1. Introduction

Microalgae are eukaryotic photosynthetic microorganisms that are adapted to live in ecologically different habitats, which results in a wide diversity of species and natural products [1,2] with pharmaceutical [3], nutraceutical [4], and cosmeceutical [5] interests. Microalgae can be cultivated in huge quantities and this advantage overcomes the bottleneck of drug discovery from marine macro-organisms and destructive collection practices. In addition, many studies have focused on optimizing the culturing conditions in order to obtain the metabolites of interest or produce them in large amount. Regarding microalgal molecular resources, various microalgal genomes have been sequenced [6]. However, there are still few genomes available compared to the huge number of existing microalgae [6]. This is due to the fact that microalgae, especially dinoflagellates, can have very large genomes, up to 112 Gbp [7]. On the contrary, several transcriptomes are available for microalgae [6]. Many of these have been sequenced thanks to the Marine Microbial Eukaryote Transcriptome Sequencing Project (MMETSP), the TARA ocean, and the Global Ocean Sampling expeditions, while others are on the way. This will also help discover the enzymes involved in the synthesis of bioactive metabolites and suggest which are the most promising species for genetic engineering manipulations in order to increase the production of specific metabolites.

At the moment, there are only a few studies on metabolic pathways involved in the synthesis of bioactive compounds from microalgae [7–11]. The aim of this paper was to investigate the presence in microalgae of enzymes involved in the synthesis of specific functional lipids with huge interest for pharmaceutical, nutraceutical, and cosmeceutical applications. In particular, microalgae have been found to contain high contents of monogalactosyldiacylglycerols (MGDGs) [12]. MGDGs are compounds of biotechnological interest because they have been investigated for the potential treatment of human pathologies such as cancer (e.g., human pancreatic cancer cell lines) [13] and inflammation [14]. A purified MGDG, extracted from spinach, is a strong DNA polymerase inhibitor and it enhances chemotherapeutic efficiency in human pancreatic cancer cell lines (BxPC-3, MIAPaCa2, and PANC-1) [15]. Regarding marine examples, two MGDGs from the diatom *Phaeodactylum tricornutum* showed *in vitro* pro-apoptotic activity on immortal mouse epithelial cell lines (W2 cells) [16].

Two different MGDGs from the green microalga (Chlorophyta) *Tetraselmis chuii* [17] and four MGDGs from *Nannochloropsis granulata* (Ochrophyta, Eustigmatophyceae) [18] showed anti-inflammatory properties. These MGDGs, extracted from both *T. chuii* and *N. granulata*, were able to reduce NO production and inducible nitric oxide synthase (iNOS) protein levels in lipopolysaccharide (LPS)-stimulated RAW264.7 macrophage cells. In addition, sulfoglycolipids, which constitute the anionic fraction of MGDGs [19], also present interesting biological properties, such as glutaminyl cyclase (QC) inhibitory activity [20] and immuno-stimulatory activity [21]. Sulfolipids extracted from the green microalgae *Tetradesmus lagerheimii* (formerly *Scenedesmus acuminatus*), *Scenedesmus producto-capitatus*, *Pectinodesmus pectinatus* (formerly *Scenedesmus pectinatus*), and *Tetradesmus wisconsinensis* are able to inhibit QC [20] (an enzyme involved in Alzheimer's disease progression [22]) and, thus, have potential as lead compounds against Alzheimer's disease. Furthermore, a synthetic sulfolipid derived from *Thalassiosira weissflogii* CCMP1336 (renamed *Conticribra weissflogii*, Bacillariophyta), named β-SQDG18, is a potent vaccine adjuvant [21]. β-SQDG18 is able to trigger a more effective immune response against cancer cells to improve dendritic cell (DC) maturation, to increase CD83-positive DC, and to stimulate the production of pro-inflammatory cytokines (IL-12 and INF-γ) [21].

The aim of the present work was to investigate the presence of genes involved in MGDG and sulfoglycolipid synthesis in microalgae (Bacillariophyta, Cercozoa, Chlorophyta, Chrysophyceae, Coccolithophyceae, Cryptophyta, Dictyochophyceae, Dinophyceae, Euglenophyceae, Eustigmatophyceae, Glaucophyceae, Pavlovophyceae, Pelagophyceae, Pinguiophyceae, Raphidophyceae, Rhodophyta, Synurophyceae, and Xanthophyceae) because little information is available. Our analyses were performed considering all the microalgal transcriptomes and metatranscriptomes available from the MMETSP project and the recent Tara Oceans and Global Ocean sampling expeditions, respectively.

Galactolipids represent up to 80% of the total lipids of the plastid membranes [23]. They can contain one or two galactose (Gal) molecules, which bond to the glycerol backbone at sn-3 position, and are called, respectively, MGDGs or digalactosyldiacylglycerols (DGDGs). The reaction required for MGDG biosynthesis is catalyzed by a MGDG synthase (MGD), which transfers the Gal moiety from UDP-Gal to diacylglycerol [24] (Figure 1a). MGDG is required in plastids biogenesis and integrity and in photosynthesis [24]. In terrestrial plants, the MGDG synthase is encoded by two types of genes, namely type-A (AtMGD1) and type-B (AtMGD2 and AtMGD3) isoforms, and enzymes are well characterized [25–30]. In microalgae, there is no information available yet.

Sulfolipids are constituent of the thylakoids in plant and algal chloroplasts [31]. Sulfoquinovose, which is the building block of sulfolipids, is also the major component of the biological sulfur cycle [32] and it is produced by photosynthetic organisms at a rate of 1010 tons per year [33]. The biosynthesis of these lipids proceeds in two reactions. The first reaction is catalysed by the UDP-sulfoquinovose synthase (SQD1), that assembles the UDP-sulfoquinovose from UDP-glucose and sulphite [34]. The second reaction is catalysed by sulfoquinovosyltransferase (SQD2), which transfers the sulfoquinovose to diacylglycerol [35]. This produces sulfoquinovosyldiacylglycerol (SQDG) (Figure 1b).

Different classes of microalgal-derived compounds have been identified and several have shown specific biological activities, such as anti-cancer [13], anti-inflammatory [14], anti-tuberculosis [36], anti-epilepsy [37], anti-microbial [38], immune-regulatory [39], anti-hypertensive [40], anti-atherosclerosis [40], and anti-osteoporosis [40] activity. The systematic studies of the genes involved in the synthesis of the bioactive compounds of interest are increasing [7,9,10] and can open new perspectives for gene-editing and boost the use of microalgae as a source of new marine natural products.

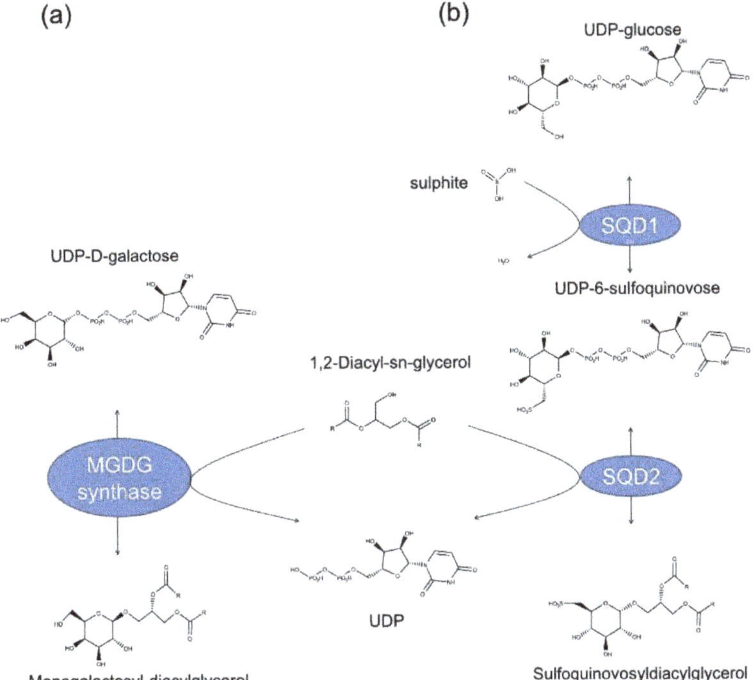

Figure 1. Enzymes responsible for (**a**) monogalactosyldiacylglycerol (i.e., MGDG synthase) and (**b**) sulfoquinovosyldiacylglycerol (i.e., UDP-sulfoquinovose synthase or SQD1 and sulfoquinovosyltransferase or SQD2) biosynthesis.

2. Results and Discussion

2.1. Identification and Taxonomic Assignation of MGD, SQD1, and SQD2 Homologous Sequences

The BLASTp search conducted in the OGA database returned 1,614 annotated eukaryotic sequences putatively attributable to MGD, 1,422 to SQD1 and 1,340 to SQD2. Of these sequences, the following were unambiguously identified as the genes of interest after Blast2GO analysis: 1154 to MGDG, 817 to SQD1, and 273 to SQD2 (Supplementary Data S1–S3, respectively).

The taxonomic pattern of these sequences, excluding the ones annotated as only as "Eukaryota" and "Stramenopiles" is illustrated in Figure 2. Of the 15 taxonomic divisions considered, 13 were found in the sequences assigned to MGD, 11 to SQD1, and 7 to SQD2 (Figure 2). In this database (OGA), we found homologs for all three genes investigated in the following microalgal taxa: Bacillariophyta, Chlorophyta, Dictyochophyceae, Coccolithophyceae, Pavlovophyceae, and Pelagophyceae. In particular, homolog sequences were particularly abundant in Bacillariophyta, Coccolithophyceae, and Pelagophyceae. For Pinguiophyceae, Raphidophyceae, and Synurophyceae, only sequences coding for MGDG were found. On the contrary, sequences assigned to Ciliophora and Eustigmatophyceae were exclusively corresponding to SQD1 and SQD2. Considering that the results are based on transcriptomic data, they can suggest the absence of a specific gene in certain taxa or simply the analyzed organisms were not expressing the specific gene at the time of sampling and fixing for the RNA-sequencing.

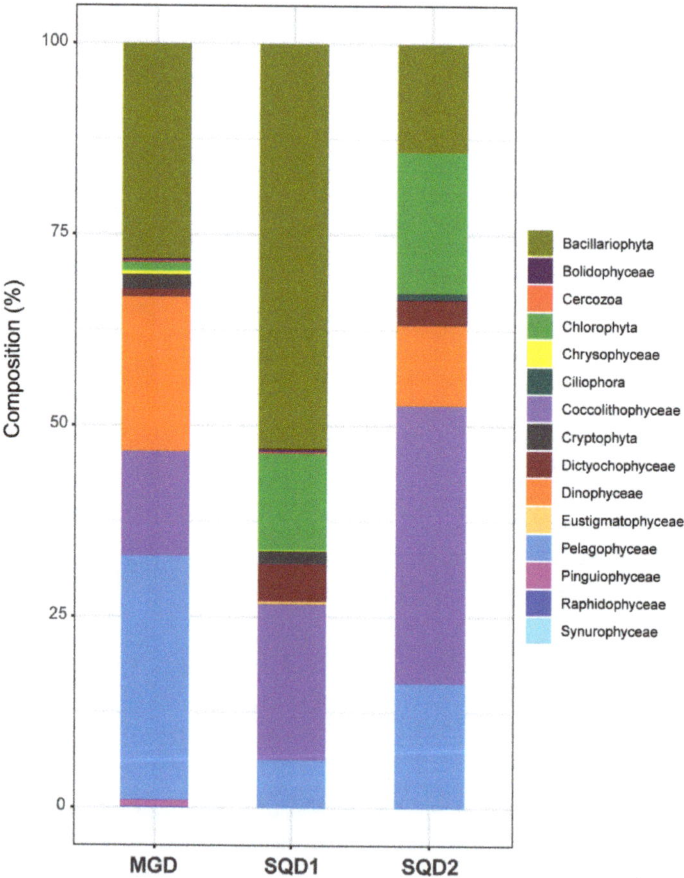

Figure 2. Taxonomic characterization and composition of the analyzed genes. MGD, SQD1, and SQD2 homologs were retrieved from Tara Oceans meta-transcriptomes. Graphical view of taxonomic composition for each gene is reported.

For MMETSP transcriptomes, we retrieved 313 sequences for MGD, 84 for SQD1, and 266 for SQD2 after performing a blastp search. After functional annotation (and removal of sequences considered not valid), 267 sequences were left for MGD, 80 for SQD1, and 121 for SQD2 (Supplementary Data S4–S6, respectively).

We reported in Table 1 the species and the strains in which MGD, SQD1, and SQD2 were found. The analyzed genes were found in 96 different species or strains across 14 different taxonomic categories. All the three genes were found in the reported taxa, with the exception of the two species of Chrysophyceae (*Dinobryon* sp. UTEXLB2267 and *Paraphysomonas imperforata* PA2, Supplementary Data S7). In addition, considering that the results are based on transcriptomic data, they can be interpreted as either the absence of that specific gene or lack of its expression in that particular condition.

Table 1. Species and strains from MMETSP transcriptomes in which MGD, SQD1, and SQD2 homologs were found. Colors refer to taxonomic ranks as in Figure 2. P = presence of validated gene homologs.

Color Legend	Taxonomic Group	Species/Strain Surveyed	Accepted Synonym	MGD	SQD1	SQD2
	Dinophyceae	*Alexandrium monilatum* CCMP3105		P	P	P
	Dinophyceae	*Alexandrium tamarense* CCMP1771		P	P	
	Dinophyceae	*Amphidinium carterae* CCMP1314		P		
	Bacillariophyta	*Amphiprora* sp.		P	P	P
	Bacillariophyta	*Amphora coffeaeformis* CCMP127	*Halamphora coffeiformis*	P	P	P
	Bacillariophyta	*Asterionellopsis glacialis* CCMP134		P	P	P
	Pelagophyceae	*Aureococcus anophagefferens* CCMP1850		P		P
	Pelagophyceae	*Aureoumbra lagunensis* CCMP1510		P	P	P
	Dinophyceae	*Azadinium spinosum* 3D9		P		
	Bacillariophyta	*Ceratium fusus* PA161109	*Tripos fusus*	P		
	Bacillariophyta	*Chaetoceros affinis* CCMP159		P	P	P
	Bacillariophyta	*Chaetoceros curvisetus*		P	P	P
	Bacillariophyta	*Chaetoceros debilis* MM31A 1		P	P	P
	Bacillariophyta	*Chaetoceros neogracile* CCMP1317		P	P	P
	Raphidophyceae	*Chattonella subsalsa* CCMP2191		P	P	P
	Coccolithophyceae	*Chrysochromulina polylepis* CCMP1757	*Prymnesium polylepis*	P	P	P
	Bacillariophyta	*Corethron pennatum* L29A3		P	P	P
	Dinophyceae	*Crypthecodinium cohnii* Seligo		P		
	Bacillariophyta	*Ditylum brightwellii* GSO103		P	P	P
	Bacillariophyta	*Ditylum brightwellii* GSO104		P	P	P
	Bacillariophyta	*Ditylum brightwellii* GSO105		P	P	P
	Chlorophyta	*Dunaliella tertiolecta* CCMP1320		P	P	P
	Dinophyceae	*Durinskia baltica* CSIRO CS 38	*Durinskia dybowskii*	P	P	P
	Coccolithophyceae	*Emiliania huxleyi* 374		P		P
	Coccolithophyceae	*Emiliania huxleyi* 379		P	P	P
	Coccolithophyceae	*Emiliania huxleyi* CCMP370		P	P	P
	Coccolithophyceae	*Emiliania huxleyi* PLYM219		P	P	P
	Euglenophyceae	*Eutreptiella gymnastica* like CCMP1594		P		P
	Bacillariophyta	*Extubocellulus spinifer* CCMP396		P	P	
	Bacillariophyta	*Fragilariopsis kerguelensis* L26 C5		P	P	P
	Bacillariophyta	*Fragilariopsis kerguelensis* L2 C3		P	P	P
	Coccolithophyceae	*Gephyrocapsa oceanica* RCC1303		P	P	P
	Dinophyceae	*Glenodinium foliaceum* CCAP1116 3	*Kryptoperidinium foliaceum*	P	P	P
	Glaucophyceae	*Gloeochaete wittrockiana* SAG46 84		P	P	P
	Cryptophyta	*Goniomonas pacifica* CCMP1869			P	
	Raphidophyceae	*Heterosigma akashiwo* CCMP2393		P	P	P
	Raphidophyceae	*Heterosigma akashiwo* CCMP3107		P	P	P
	Raphidophyceae	*Heterosigma akashiwo* CCMP452		P	P	P
	Raphidophyceae	*Heterosigma akashiwo* NB		P	P	P
	Coccolithophyceae	*Isochrysis galbana* CCMP1323		P	P	P
	Coccolithophyceae	*Isochrysis* sp. CCMP1244		P	P	P
	Coccolithophyceae	*Isochrysis* sp. CCMP1324		P	P	P
	Dinophyceae	*Karenia brevis* CCMP2229		P		

Table 1. *Cont.*

Color Legend	Taxonomic Group	Species/Strain Surveyed	Accepted Synonym	MGD	SQD1	SQD2
	Dinophyceae	*Karenia brevis* SP1		P		
	Dinophyceae	*Karenia brevis* SP3		P		
	Dinophyceae	*Karenia brevis* Wilson		P		P
	Dinophyceae	*Karlodinium micrum* CCMP2283	*Karlodinium veneficum*	P		P
	Dinophyceae	*Kryptoperidinium foliaceum* CCMP1326		P	P	P
	Dinophyceae	*Lingulodinium polyedra* CCMP1738		P	P	
	Cercozoa	*Lotharella globosa* CCCM811		P	P	P
	Chlorophyta	*Micromonas* sp. CCMP2099		P		
	Chlorophyta	*Micromonas* sp. NEPCC29		P	P	
	Chlorophyta	*Micromonas* sp. RCC472		P	P	P
	Bacillariophyta	*Nitzschia punctata* CCMP561	*Tryblionella punctata*	P	P	P
	Chrysophyceae	*Ochromonas* sp. CCMP1393		P	P	P
	Dinophyceae	*Oxyrrhis marina*		P		P
	Dinophyceae	*Oxyrrhis marina* LB1974				P
	Pavlovophyceae	*Pavlova* sp. CCMP459		P	P	P
	Pelagophyceae	*Pelagococcus subviridis* CCMP1429		P		
	Pelagophyceae	*Pelagomonas calceolata* CCMP1756		P	P	P
	Dinophyceae	*Peridinium aciculiferum* PAER 2	*Apocalathium aciculiferum*	P		
	Chlorophyta	*Picocystis salinarum* CCMP1897			P	
	Coccolithophyceae	*Pleurochrysis carterae* CCMP645	*Chrysotila carterae*	P		P
	Bacillariophyta	*Proboscia alata* PI D3		P	P	P
	Dinophyceae	*Prorocentrum minimum* CCMP1329	*Prorocentrum cordatum*	P	P	
	Dinophyceae	*Prorocentrum minimum* CCMP2233	*Prorocentrum cordatum*	P	P	
	Coccolithophyceae	*Prymnesium parvum* Texoma1		P	P	P
	Bacillariophyta	*Pseudo-nitzschia australis* 10249 10 AB		P	P	P
	Bacillariophyta	*Pseudo-nitzschia fradulenta* WWA7		P	P	P
	Dictyochophyceae	*Pseudopedinella elastica* CCMP716		P	P	P
	Dictyochophyceae	*Pteridomonas danica* PT		P	P	P
	Chlorophyta	*Pyramimonas parkeae* CCMP726		P	P	P
	Rhodophyta	*Rhodella maculata* CCMP736	*Rhodella violacea*	P	P	
	Cryptophyta	*Rhodomonas* sp. CCMP768		P	P	P
	Dinophyceae	*Scrippsiella hangoei* SHTV5	*Apocalathium malmogiense*	P		
	Dinophyceae	*Scrippsiella hangoei* like SHHI 4		P		
	Dinophyceae	*Scrippsiella trochoidea* CCMP3099		P		
	Bacillariophyta	*Skeletonema dohrnii* SkelB		P	P	P
	Bacillariophyta	*Skeletonema marinoi* SkelA		P	P	P
	Bacillariophyta	*Skeletonema menzelii* CCMP793		P	P	P
	Dinophyceae	*Symbiodinium* sp. C1		P		
	Dinophyceae	*Symbiodinium* sp. CCMP2430		P		
	Dinophyceae	*Symbiodinium* sp. Mp		P		
	Chlorophyta	*Tetraselmis striata* LANL1001		P	P	P
	Bacillariophyta	*Thalassionema nitzschioides* L26 B		P		P
	Bacillariophyta	*Thalassiosira antarctica* CCMP982		P	P	P
	Bacillariophyta	*Thalassiosira gravida* GMp14c1		P	P	P
	Bacillariophyta	*Thalassiosira miniscula* CCMP1093		P	P	P
	Bacillariophyta	*Thalassiosira oceanica* CCMP1005		P	P	P
	Bacillariophyta	*Thalassiosira rotula* CCMP3096	*Thalassiosira gravida*	P	P	P
	Bacillariophyta	*Thalassiosira rotula* GSO102	*Thalassiosira gravida*	P	P	P
	Bacillariophyta	*Thalassiosira weissflogii* CCMP1010	*Conticribra weissflogii*	P	P	
	Bacillariophyta	*Thalassiosira weissflogii* CCMP1336	*Conticribra weissflogii*	P	P	P
	Bacillariophyta	*Thalassiothrix antarctica* L6 D1		P	P	
	Xanthophyceae	*Vaucheria litorea* CCMP2940		P	P	P

2.2. Final Dataset and Phylogenetic Inferences

The final alignments consisted of 649 sequences for MGD, 521 for SQD1, and 244 for SQD2 (Supplementary Data S8–S10, respectively), from both the OGA and MMETSP database. The length of each dataset after trimming of poorly aligned regions was as follows: 377 bp for MGD, 198 bp for SQD1, and 295 bp for SQD2.

The MGD phylogenetic tree showed that most of the taxa here investigated contained paralog copies (homolog copies resulting from duplication events) of monogalactosyldiacylglycerol synthase gene (Figure 3). These paralogs generally occurred in two copies, resulting in two distinct and highly supported clades for most taxa (red circles, Figure 3, Supplementary File S1). A few taxa (e.g., Coccolithophyceae, Cryptophyta, and Xantophyceae) presented only one MGD copy (and, therefore, a single group of sequences), while others (e.g., Dinophyceae) presented several MGD paralogs (Figure 3 or Supplementary File S1).

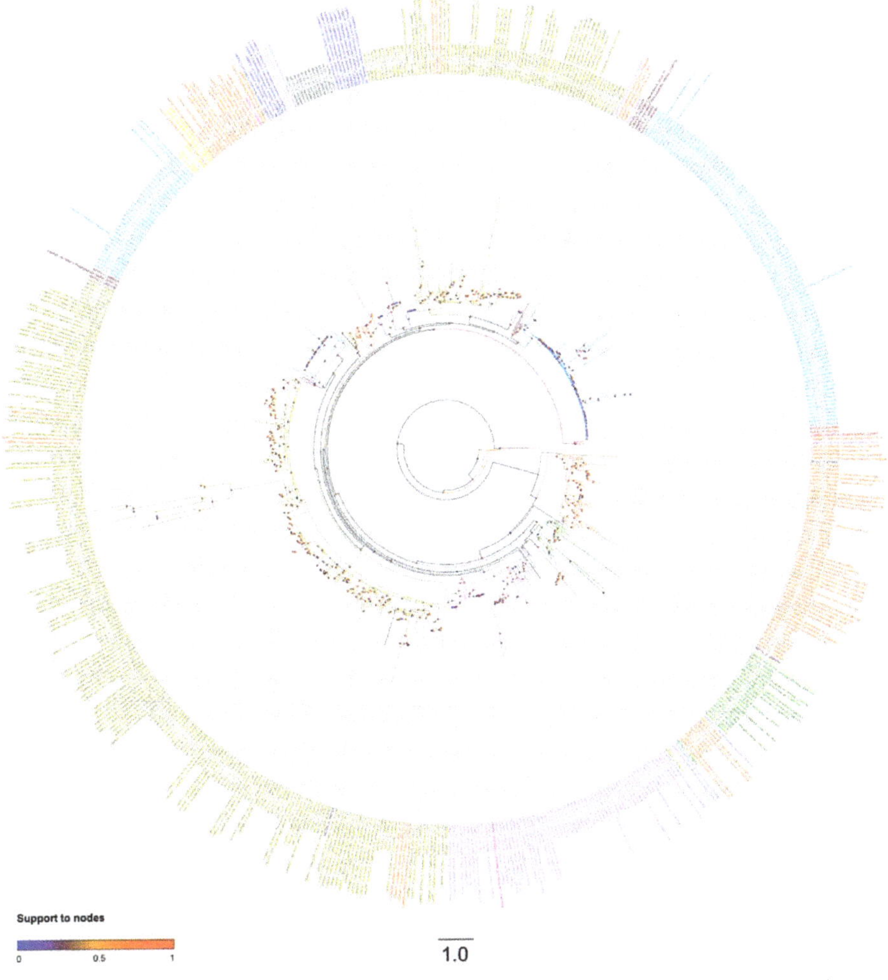

Figure 3. MGD unrooted phylogenetic tree. Colored circles at the base of each node refer to branch support after aLRT SH-like test. Colors of taxa refer to taxonomic groups as in Figure 2 and Table 1.

For the SQD1 gene, the phylogenetic tree showed that all the sequences belonging to the same taxon formed a highly supported monophyletic group in most species (Figure 4, Supplementary File S2). Among the exceptions, there are the diatoms (Bacillariphyta) where, beside a clade containing the most diatom sequences retrieved from OGA and MMETSP, there are a few others interspersed across the tree (Figure 4). One of these only contains sequences of *Thalassiosira rotula* (synonym of *T. gravida*) strain CCMP1093 from MMETSP. The others contain small groups of sequences from OGA (Figure 4, Supplementary File S2). Similarly, SQD1 dinoflagellates (Dinophyceae) homologs do not form a monophyletic group in the tree. This could be due to the high complexity of the plastid evolution in Dinophyceae. Dinophyceae acquired plastid with four different endosymbiotic events. [41]. In a secondary endosymbiotic event, Dinophyceae acquired plastid by serial endosymbiosis with green microalgae (Chlorophyta) [42]. In the tertiary endosymbiotic event, Dinophyceae acquired tertiary plastids by endosymbiosis with Cryptophyceae-Cryptomonads, Haptophyta, and Bacillariphyta [41].

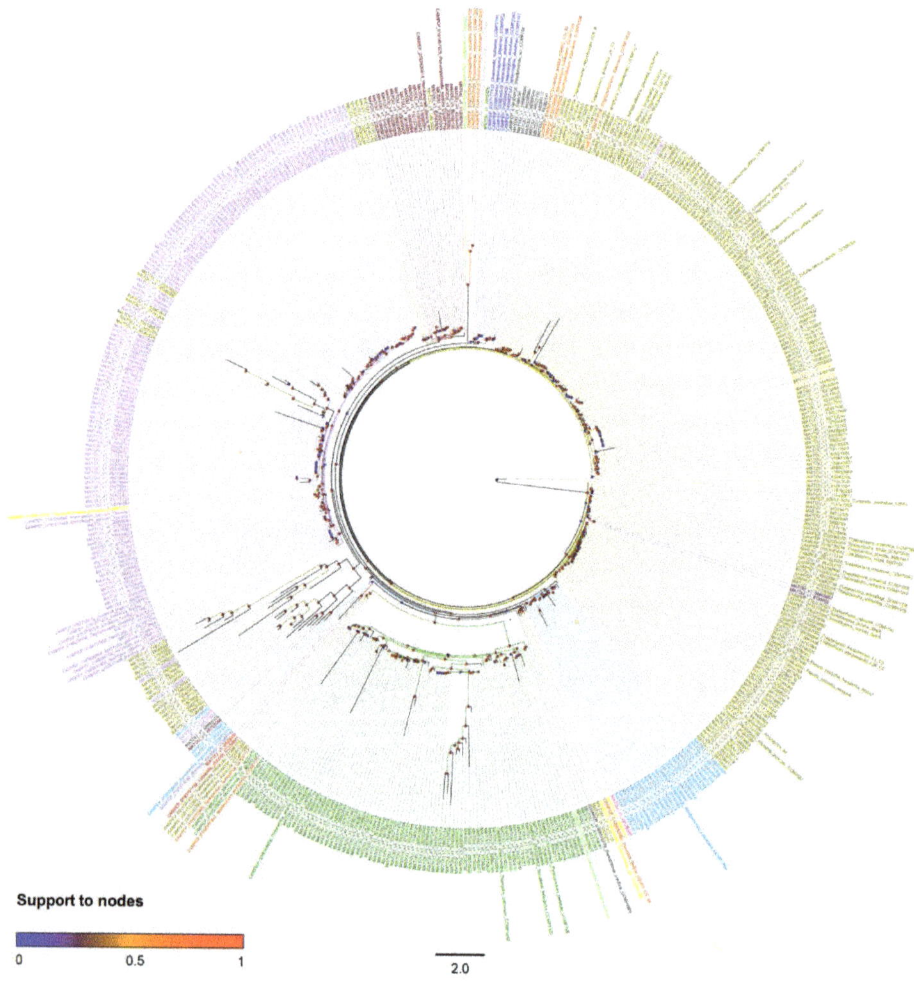

Figure 4. SQD1 unrooted the phylogenetic tree. Colored circles at the base of each node refer to branch support after an aLRT SH-like test. Colors of taxa refer to taxonomic groups in Figure 2 and Table 1.

For the SQD2 gene, all the sequences from the same taxonomic group formed monophyletic groups except for the Dinophyceae (Figure 5, Supplementary File S3). All the dinoflagellate sequences from OGA and MMETSP were interspersed across the tree with high support, close to Bacillariophyta, Chlorophyta, Dictyocophyceae, and Coccolithophyceae.

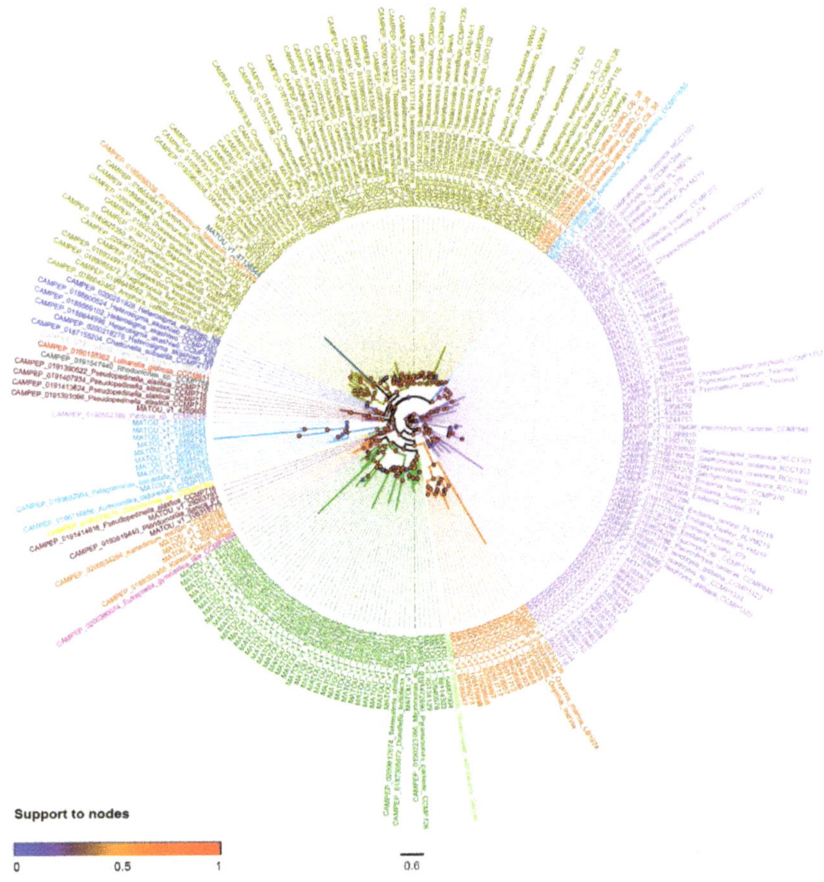

Figure 5. SQD2 unrooted phylogenetic tree. Colored circles at the base of each node refer to branch support after the aLRT SH-like test. Colors of taxa refer to taxonomic groups in Figure 2 and Table 1.

2.3. Structural Details of MDGs, SQD1, and SQD2 from Thalassiosira Weissflogii CCMP1336 (Conticriba weissflogii)

We built *in silico* models for MGDs, SQD1, and SQD2 proteins from the diatom *T. weissflogii* CCMP1336 (synonym of *Conticriba weissflogii*). *T. weissflogii* (*C. weissflogii*) CCMP1336 has been selected because it is known to have immunostimulatory activity, and both galactolipids and sulfolipids have been found to play a key role in this bioactivity [39,43]. 3D *in silico* models (Figure 6) were generated using the amino acid sequences of MGD (CAMPEP_0193073380, CAMPEP_0193064960 and CAMPEP_0193062160) (Figure 6a–c), SQD1 (CAMPEP_0193062736) (Figure 6d), and SQD2 (CAMPEP_0193058822) (Figure 6e), obtained from the MMETSP database. Phyre2 results were summarized in Table 2. The analyses pointed out the homology between *T. weissflogii* (*C. weissflogii*) and *Arabidopsis thaliana* proteins. Structural prediction of MGDs (CAMPEP_0193073380, Supplementary Data S11, CAMPEP_0193064960 Supplementary Data S12, and CAMPEP_0193062160, Supplementary

Data S13) found high similarity with the 3D structure of MGD1 from *A. thaliana*. The percentage of identity between *T. weissflogii* (*C. weissflogii*) and *A. thaliana* MGD1 is higher than 40%. Similar results were obtained for structural prediction of SQD1 from *T. weissflogii* (*C. weissflogii*) (Supplementary Data S14) with 45% identity with the 3D structure of SQD1 from *A. thaliana*.

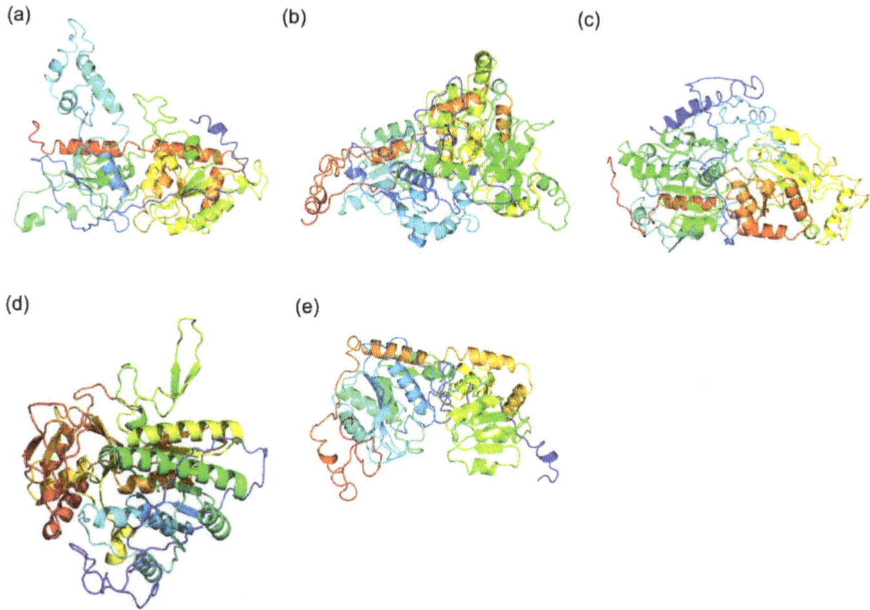

Figure 6. *In silico* model generated by Phyre2 for: (**a**) MGD CAMPEP_0193073380, (**b**) MGD CAMPEP_0193064960, (**c**) MGD CAMPEP_0193062160, (**d**) SQD1 CAMPEP_0193062736, and (**e**) SQD2 CAMPEP_0193058822.

Table 2. Report of Phyre2 analysis. We report the template (protein of known structure used for the prediction analysis) and its protein data bank (PDB) code, confidence (probability that the sequence and template are homologous), and percent id (percent of identity).

Thalassiosira (*Conticribra*) *weissflogii* CCMP1336 Proteins	Template (PDB Code)	Confidence	% id
MGD CAMPEP 0193073380	MGD1 from *A. thaliana* (4X1T)	100	46
MGD CAMPEP 0193064960	MGD1 from *A. thaliana* (4X1T)	100	47
MGD CAMPEP 0193062160	MGD1 from *A. thaliana* (4X1T)	100	40
SQD1 CAMPEP_0193062736	SQD1 from *A. thaliana* (1I24)	100	45
SQD2 CAMPEP_0193058822	SUS1 from *A. thaliana* (3S29)	100	18

Structural prediction for SQD2 (Supplementary Data S15) found a similarity with the 3D structure of sucrose synthase-1 (SUS1) from *A. thaliana*. The percentage of identity between SQD2 and SUS1 is 18%, which is lower than the percent of identity of SQD1 and MDGs. SUS1 from *A. thaliana* catalyzed a reversible sucrose synthesis. It transfers glucose moiety from UDP to fructose and it has been found to form a complex with both UDP–glucose and UDP–fructose [44]. These findings may indicate the presence of a low specificity ligand-binding site in the SQD2 enzyme.

In order to evaluate whether predicted structures preserved the functional active sites of MGDs, SQD1, and SQD2, they were also analyzed by using the COACH server. MGD analysis was performed for all the structures obtained from Phyre2. The analysis pointed out the presence of a UDP binding pocket for the three MGDs. MGD CAMPEP_0193073380 (Figure 7a) binding site prediction had a

C-score 0.32 and the presence of 17 residues in the binding site (i.e., H69, A71, R263, G312, G313, V350, G352, F379, V380, M383, Y396, G398, P399, G400, T401, E40, and E422). MGD CAMPEP_0193064960 (Figure 7b) binding site prediction had a low C-score value (0.18) and 17 amino acid residues were involved in the binding site (i.e., H89, A91, R260, G305, G306, V335, G337, F528, V529, M532, K545, G547, P548, G549, T550, E553, and E571). MGD CAMPEP_0193062160 (Figure 7c) binding site prediction had a low C-score (0.18) and 17 residues involved (i.e. H262, A264, R434, G477, G478, V507, G509, F586, V587, M590, K603, G605, P606, G607, T608, E611, and E629). COACH prediction analysis highlighted the same 17 amino acid residues involved in ligand-binding site of the three MDGs. Moreover, clustal omega analysis including MGD1 from *A. thaliana* indicated that these 17 residues were located in highly conserved domains (Figure 7d) and corresponded to the active site of MGD1 from *A. thaliana* [45] with the exception of the alanine residue.

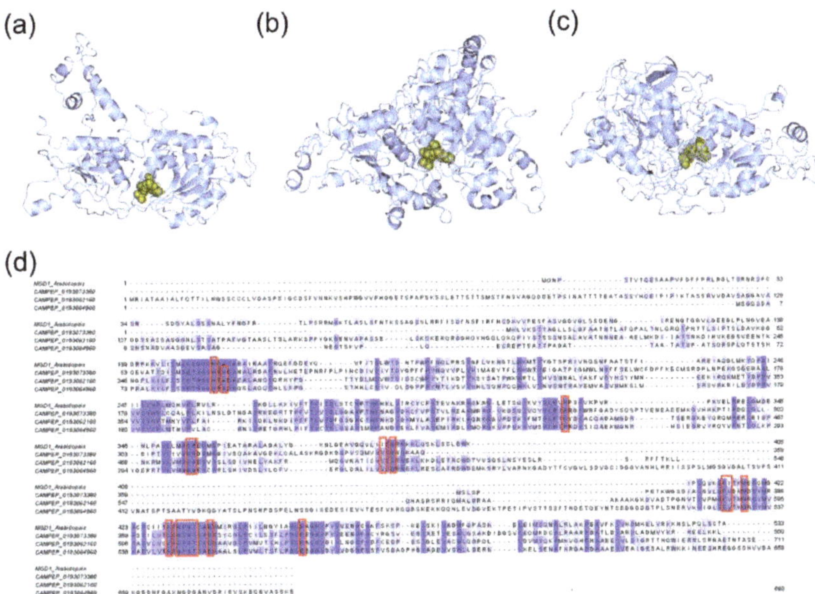

Figure 7. Protein-ligand binding site prediction by the COACH server for MDGs from *Thalassiosira* (*Conticribra*) *weissflogi* CCMP1336. Prediction of binding sites for the three MGDs: (**a**) MGD CAMPEP_0193073380–UDP complex, (**b**) MGD CAMPEP_0193064960–UDP complex, and (**c**) MGD CAMPEP_0193062160–UDP complex. (**d**) Among the Clustal Omega result of MDG amino acid sequences, the red boxes indicated the conserved amino acid residues involved in binding-pockets.

SQD1 analysis indicated the presence of a NAD-binding pocket (Figure 8a) with a C-score of 0.85. The following residues were reported to be involved in the NAD-binding pocket structure: G74, D76, G77, F78, C79, D98, N99, S101, R102, L140, D141, V142, F164, A165, E166, R168, A169, N186, L210, G211, T212, Y252, K256, Q279, G280, I281, V282, and Y306. The presence of a UDP–6–sulfoquinovose (USQ) binding pocket (Figure 8b) was predicted with a C-score of 0.27. The residues involved in USQ-binding pocket were R168, A170, T212, M213, G214, Y252, H253, Q279, G280, I281, T308, V309, R312, T324, Y326, Q331, R333, V371, R393, and E395. Most of them were involved in USQ binding (Figure 8c). The ligand binding pocket structure prediction, such as the predicted structure, was in line with the structure and function of SQD1 from *A. thaliana* [34,46].

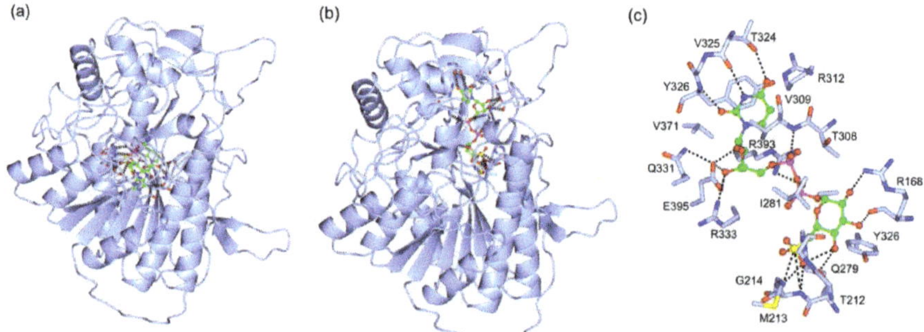

Figure 8. Protein-ligand binding site prediction by COACH server of SQD1 from *Thalassiosira* (*Conticribra*) *weissflogii* CCMP1336. (**a**) Prediction of binging site of the complex SQD1–NAD. (**b**) Prediction of binding site of the complex SQD1–USQ. (**c**) Structure of USQ binding-pocket and the specific interaction between USQ and highlighted amino acid residues of SQD1.

SQD2 analysis indicated the presence of a UDP-biding pocket (Figure 9a) (C-score 0.28), which involved the residues S82, G83, N86, F256, V282, G283, R284, K289, E309, Q332, L333, L338, E355, G358, F359, V360, and E363. Moreover, COACH analyses pointed out the presence of a N-Acetylglucosamine (NAG)–binding pocket (Figure 9b) (C-score 0.33) that involved the residues G83, Y84, R87, H190, T191, K249, S354, E355, T356, L357, and G358. These data confirmed the presence of a low specificity ligand binding pocket, and suggest that SQD2 from diatoms could be involved in the synthesis of other diacylglycerols such as glucuronosyldiacylglycerol (GlcADG) in *Arabidopsis* [47] or flavonoid glycosylation in *Oryza sativa* [48,49].

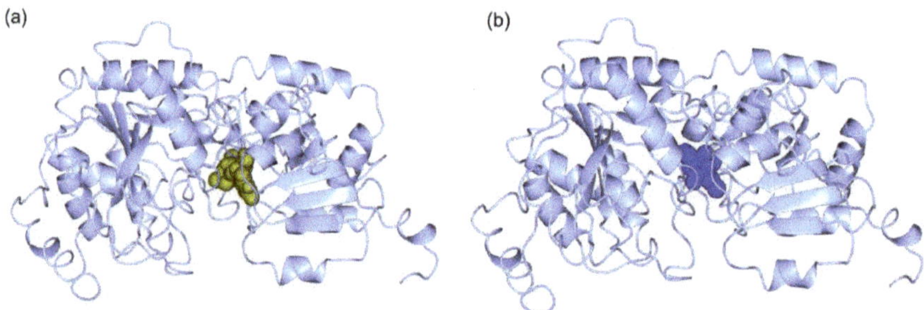

Figure 9. Protein-ligand binding site prediction by COACH server of SQD2 from *Thalassiosira* (*Conticribra*) *weissflogii* CCMP1336. (**a**) Prediction of binding site of the complex SQD2–UDP. (**b**) prediction of binding site of the complex SQD2–NAG.

3. Materials and Methods

3.1. Identification of MGD, SQD1, and SQD2 Homologous Sequences

Since our primary interest was to ascertain the occurrence of MGD, SQD1, and SQD2 in microalgae, we used the sequences of MGD (accession number XP002181685), SQD1 (accession number XP002185968), and SQD2 (accession number XP002185276) from the diatom *Phaeodactylum tricornutum* as queries for a BLAST [50] search against the Ocean Global Atlas (OGA, [51]) database (http://tara-oceans.mio.osupytheas.fr/oceangene-atlas/), and several protist transcriptomes from the Marine Microbial Eukaryotic Transcriptome Sequencing Project (MMETSP, [52]) available at https://zenodo.org/record/12125852585. The OGA database contains a collection of more than 116 million eukaryote and 40 million prokaryotic genes gathered during the *Tara* Oceans [53,54] and the Global Ocean Sampling [55] expeditions. Instead, MMETSP contains the transcriptomes of some of the most abundant and ecologically significant microbial eukaryotes in the oceans. To retrieve homologs, we used the blastp algorithm against the metagenome/metatranscriptome and transcriptome databases contained in OGA and the MMETSP, setting the expect threshold to 1E-10. For OGA, after ascertaining that the sequences obtained from metagenomes and metatranscriptomes were identical, we only used one dataset (metatranscriptomes) for further analyses. Since a sequence-based homology search could have recovered different genes with similar functions, we used Blast2GO [56] to obtain a functional annotation of the homologs retrieved. We used the default settings (i.e., blastx program, using the nr BLAST database and with a BLAST expectation value of 1.0E-3) for the analysis. We considered valid Blast2GO annotations containing the following names: 1,2-diacylglycerol 3-beta-galactosyltransferase and monogalactosyldiacylglycerol synthase for MGD, sulfolipid biosynthesis protein, sulfoquinovosyldiacylglycerol synthesis protein, UDP sulfoquinovose synthase and uridine 5′-diphosphate-sulfoquinovose synthase for SQD1, sulfoquinovosyl transferase SQD2, sulfoquinovosyldiacylglycerol 2, and UDP-sulfoquinovose: DAG sulfoquinovosyltransferase for SQD2. All the sequences identified as only "predicted or hypothetical protein" without specification were *a priori* discarded.

3.2. Taxonomic Overview

All the sequences from OGA that passed the Blast2GO analysis with the criterions specified above were annotated using the annotation file generated during homolog retrieval. We removed all the sequences without a taxonomic annotation, annotated only as "Eukaryota" and, whenever applicable, with a generic annotation that did not allow to discriminate such sequences from others of lower taxonomic rank (e.g., "Stramenopiles"). All the taxa in which the MGD, SQD1, and SQD2 genes were found and were organized into a table. The abundance of such sequences were plotted as histograms using the R [57] working packages *scales* [58] and *ggplot2* [59]. For the homologs retrieved from the MMETSP transcriptomes, we generated a table illustrating the species surveyed and the genes of interest that were found in each of them.

3.3. Sequence Alignment and Phylogenetic Inference

The sequences of each gene from OGA and MMETSP that passed Blast2GO annotation and with a length of at least 200 aa were aligned using COBALT [60] available at https://www.ncbi.nlm.nih.gov/tools/cobalt/). Unlike other common software used for protein alignments (e.g., ClustalW, MAFFT, MUSCLE, ProbCons) that only use sequence information, COBALT also integrates the information of protein-motif regular expressions (PROSITE database) and of conserved protein domains (NCBI CDD database). In doing so, COBALT has a better chance of producing a biologically meaningful multiple alignment compared to tools that do not utilize this information 60]. Poorly aligned regions from each alignment were removed with trimAl v1.2 [61] in order to increase the quality of subsequent phylogenetic analyses. We used the *automated1* option to find the most appropriate mode to trim the alignments (use of gaps or similarity scores) depending on the alignment characteristics.

In order to infer reliable phylogenetic trees, all the sequences that, after trimming, were shorter than one quarter (around 60–70 aa) of the final length of the alignment were removed, unless longer sequences were not available for that particular taxon. Maximum likelihood phylogenetic trees were inferred using PhyML [62] using the LG substitution model [63], which turned out to be the best evolution model for the three genes investigated according to the Akaike Information Criterion implemented in SMS [64]. Support to nodes was calculated using the Shimodaira-Hasegawa-like (aLRT SH-like) procedure [65]. Trees were visualised and graphically edited in FigTree v1.4.3 (http://tree.bio.ed.ac.uk/software/figtree/).

3.4. In Silico Protein Model

The three-dimensional (3D) *in silico* models of MGD, SQD1, and SQD2 proteins from *Thalassiosira weissflogii* CCMP1336 (currently regarded as synonym of *Conticribra weissflogii*, Bacillariophyta) were generated using Phyre2 server (http://www.sbg.bio.ic.ac.uk/~{}phyre2/html/page.cgi?id=index) [66]. Protein ligand-binding site predictions were performed using COACH analyses from the Zhang Lab server. COACH generated ligand-binding site predictions and a confidence score (C-score) of the prediction. The C-score ranged from 0 to 1. A higher score indicates a reliable prediction (https://zhanglab.ccmb.med.umich.edu/COACH/) [67,68]. Pictures were obtained using CCP4MG, version 2.10.11 obtained by http://www.ccp4.ac.uk/MG/download/ [69].

4. Conclusions

This study analyzes the presence of MGD, SQD1, and SQD2 in microalgae and gives a broad overview of their presence among different microalgal classes. It has been shown that, within each taxa, some species do not express all the sequences, which suggests the absence of a specific gene. Despite the fact that several species have been found to express MGDs, SQD1, and SQD2 (Table 1), only a relatively small number of active microalgae have been studied and reported to possess anti-inflammatory/immunomodulatory activities, including the diatoms *Attheya longicornis*, *Cylindrotheca closterium*, *Trieres mobiliensis* (formerly *Odontella mobiliensis*), *Phaeodactylum tricornutum*, *Porosira glacialis*, *Pseudo-nitzschia pseudodelicatissima*, and *Thalassiosira* (*Conticribra*) *weissflogii*, and the flagellates *Amphidinium carterae* (Dinophyceae), *Edaphochlamys debaryana* (formerly *Chlamydomonas debaryana*), *Chloroidium saccharophilum* (formerly *Chlorella ovalis*), *Dunaliella salina* (formerly *Dunaliella bardawil*) (Chlorophyceae), *Nannochloropsis granulata*, *Nannochloropsis oculata* (Eustigmatophyceae), *Diacronema lutheri* (formerly *Pavlova lutheri*, Pavlovophyceae), *Tetraselmis chuii*, and *Tetraselmis suecica* (Chlorophyta, Chlorodendrophyceae) [39]. Our study can direct and accelerate the discovery of new bioactive species and give new insights on enzyme discovery from microalgae with biotechnological applications.

3D *in silico* prediction analyses indicated that the enzymes, maintaining conserved domains, could be effectively involved in the synthesis of compounds with known anticancer and immune-modulatory activities, such as MGDGs and SQDGs. This approach can give preliminary information for the selection of specific microalgal species for drug discovery and for genetic engineering approaches in order to produce huge amounts of bioactive compounds of pharmaceutical interest.

Supplementary Materials: The following are available online at http://www.mdpi.com/1660-3397/18/5/237/s1. **Data S1**: Blast2GO analysis of MGD genes. BLASTp search conducted in the OGA database highlighted 1614 annotated eukaryotic sequences putatively attributable to MGD. Of these sequences, 1154 were identified as the gene of interest. **Data S2**: Blast2GO analysis of SQD1 genes. BLASTp search conducted in the OGA database highlighted 1422 annotated eukaryotic sequences putatively attributable to SQD1. Of these sequences, 817 were identified as the gene as the gene of interest. **Data S3**: Blast2GO analysis of SQD2 genes. BLASTp search conducted in the OGA database highlighted 1340 annotated eukaryotic sequences putatively attributable to SQD2. Of these sequences, 273 were identified as the gene of interest. **Data S4**: MGD amino acid sequences from MMETSP transcriptomes. After functional annotation, 267 sequences were left for MGD. **Data S5**: SQD1 amino acid sequences from MMETSP transcriptomes. After functional annotation, 84 sequences were left for MGD. **Data S6**: SQD2 amino acid sequences from MMETSP transcriptomes. After functional annotation, 121 sequences were left for SQD2. **Data S7**: Original data for Table 1. All the species and the strains were analyzed from MMETSP. **Data S8**: Alignments for MGD amino acid sequences. The alignments were performed using 649 amino acid sequences for MGD. **Data S9**: Alignments for SQD1 amino acid sequences. The alignments were performed using 521 amino acid sequences for SQD1. **Data S10**: Alignments for SQD2 amino acid sequences. The alignments were performed using 295 amino acid sequences for SQD2. **Data S11**: PDB file of MGD CAMPEP_0193073380 predicted structure. *In silico* model generated by Phyre2 for MGD CAMPEP_0193073380. **Data S12**: PDB file of MGD CAMPEP_0193064960 predicted structure. *In silico* model generated by Phyre2 for MGD CAMPEP_0193064960. **Data S13**: PDB file of MGD CAMPEP_0193062160 predicted structure. *In silico* model generated by Phyre2 for MGD CAMPEP_0193062160. **Data S14**: PDB file of SQD1 CAMPEP_0193062736 predicted structure. *In silico* model generated by Phyre2 for SQD1 CAMPEP_0193062736. Additional file 15: Data S15: PDB File of SQD2 CAMPEP_0193058822 predicted structure. *In silico* model generated by Phyre2 for SQD2 CAMPEP_0193058822. **File S1**: PDF file of MGD unrooted phylogenetic tree. Colored circles at the base of each node refer to branch support after aLRT SH-like test. Colors of taxa refer to taxonomic groups as in Table 1. **File S2**: PDF file of SQD1 unrooted phylogenetic tree. Colored circles at the base of each node refer to branch support after aLRT SH-like test. Colors of taxa refer to taxonomic groups in Table 1. **File S3**: PDF file of SQD2 unrooted phylogenetic tree. Colored circles at the base of each node refer to branch support after the aLRT SH-like test. Colors of taxa refer to taxonomic groups as in Table 1.

Author Contributions: Conceptualization, G.R. and C.L. Methodology and formal analysis, G.R., D.D.L., and C.L. Supervision, C.L., G.R., D.D.L., and C.L. co-wrote the paper. All authors have read and agreed to the published version of the manuscript.

Funding: The "Antitumor Drugs and Vaccines from the Sea (ADViSE)" project (PG/2018/0494374) funded this research.

Conflicts of Interest: The authors declare no conflict of interest.

References

1. Plaza, M.; Herrero, M.; Cifuentes, A.; Ibanez, E. Innovative natural functional ingredients from microalgae. *J. Agric. Food Chem.* **2009**, *57*, 7159–7170. [CrossRef] [PubMed]

2. Jaspars, M.; De Pascale, D.; Andersen, J.H.; Reyes, F.; Crawford, A.D.; Ianora, A. The marine biodiscovery pipeline and ocean medicines of tomorrow. *J. Mar. Biol. Assoc. UK* **2016**, *96*, 151–158. [CrossRef]

3. Mimouni, V.; Ulmann, L.; Pasquet, V.; Mathieu, M.; Picot, L.; Bougaran, G.; Cadoret, J.P.; Morant-Manceau, A.; Schoefs, B. The potential of microalgae for the production of bioactive molecules of pharmaceutical interest. *Curr. Pharm. Biotechnol.* **2012**, *13*, 2733–2750. [CrossRef] [PubMed]

4. Koyande, A.K.; Chew, K.W.; Rambabu, K.; Tao, Y.; Chu, D.-T.; Show, P.-L. Microalgae: A potential alternative to health supplementation for humans. *Food Sci. Hum. Wellness* **2019**, *8*, 16–24. [CrossRef]

5. Mourelle, M.L.; Gómez, C.P.; Legido, J.L. The potential use of marine microalgae and cyanobacteria in cosmetics and thalassotherapy. *Cosmetics* **2017**, *4*, 46. [CrossRef]

6. Lauritano, C.; Ferrante, M.I.; Rogato, A. Marine natural products from microalgae: An -omics overview. *Mar. Drugs* **2019**, *17*, 269. [CrossRef]

7. Lauritano, C.; De Luca, D.; Ferrarini, A.; Avanzato, C.; Minio, A.; Esposito, F.; Ianora, A. De novo transcriptome of the cosmopolitan dinoflagellate *Amphidinium carterae* to identify enzymes with biotechnological potential. *Sci. Rep.* **2017**, *7*, 11701. [CrossRef]

8. Di Dato, V.; Di Costanzo, F.; Barbarinaldi, R.; Perna, A.; Ianora, A.; Romano, G. Unveiling the presence of biosynthetic pathways for bioactive compounds in the *Thalassiosira rotula* transcriptome. *Sci. Rep.* **2019**, *9*, 9893. [CrossRef]

9. Lauritano, C.; De Luca, D.; Amoroso, M.; Benfatto, S.; Maestri, S.; Racioppi, C.; Esposito, F.; Ianora, A. New molecular insights on the response of the green alga *Tetraselmis suecica* to nitrogen starvation. *Sci. Rep.* **2019**, *9*, 3336. [CrossRef]

10. Elagoz, A.M.; Ambrosino, L.; Lauritano, C. De novo transcriptome of the diatom *Cylindrotheca closterium* identifies genes involved in the metabolism of anti-inflammatory compounds. *Sci. Rep.* **2020**, *10*, 4138. [CrossRef]

11. Vingiani, G.M.; De Luca, P.; Ianora, A.; Dobson, A.D.W.; Lauritano, C. Microalgal enzymes with biotechnological applications. *Mar. Drugs* **2019**, *17*, 459. [CrossRef] [PubMed]

12. Junpeng, J.; Xupeng, C. Monogalactosyldiacylglycerols with High PUFA Content From Microalgae for Value-Added Products. *Appl. Biochem. Biotechnol.* **2019**. [CrossRef] [PubMed]

13. Martinez Andrade, K.A.; Lauritano, C.; Romano, G.; Ianora, A. Marine Microalgae with Anti-Cancer Properties. *Mar. Drugs* **2018**, *16*, 165. [CrossRef] [PubMed]

14. Da Costa, E.; Silva, J.; Mendonça, S.H.; Abreu, M.H.; Domingues, M.R. Lipidomic approaches towards deciphering glycolipids from microalgae as a reservoir of bioactive lipids. *Mar. Drugs* **2016**, *14*, 101. [CrossRef] [PubMed]

15. Akasaka, H.; Sasaki, R.; Yoshida, K.; Takayama, I.; Yamaguchi, T.; Yoshida, H.; Mizushina, Y. Monogalactosyl diacylglycerol, a replicative DNA polymerase inhibitor, from spinach enhances the anti-cell proliferation effect of gemcitabine in human pancreatic cancer cells. *Biochim. Biophys. Acta Gen. Subj.* **2013**, *1830*, 2517–2525. [CrossRef]

16. Andrianasolo, E.H.; Haramaty, L.; Vardi, A.; White, E.; Lutz, R.; Falkowski, P. Apoptosis-inducing galactolipids from a cultured marine diatom, *Phaeodactylum tricornutum*. *J. Nat. Prod.* **2008**, *71*, 1197–1201. [CrossRef]

17. Banskota, A.H.; Gallant, P.; Stefanova, R.; Melanson, R.; Oleary, S.J.B. Monogalactosyldiacylglycerols, potent nitric oxide inhibitors from the marine microalga *Tetraselmis chui*. *Nat. Prod. Res.* **2013**, *23*, 1084–1090. [CrossRef]

18. Banskota, A.H.; Stefanova, R.; Gallant, P.; McGinn, P.J. Mono- and digalactosyldiacylglycerols: Potent nitric oxide inhibitors from the marine microalga *Nannochloropsis granulata*. *J. Appl. Phycol.* **2013**, *27*, 349–357. [CrossRef]

19. Kobayashi, K. Role of membrane glycerolipids in photosynthesis, thylakoid biogenesis and chloroplast development. *J. Plant. Res.* **2016**, *129*, 565–580. [CrossRef]

20. Hielscher-Michael, S.; Griehl, C.; Buchholz, M.; Demuth, H.U.; Arnold, N.; Wessjohann, L.A. Natural Products from Microalgae with Potential against Alzheimer's Disease: Sulfolipids Are Potent Glutaminyl Cyclase Inhibitors. *Mar. Drugs* **2016**, *14*, 203. [CrossRef]

21. Manzo, E.; Cutignano, A.; Pagano, D.; Gallo, C.; Barra, G.; Nuzzo, G.; Sansone, C.; Ianora, A.; Urbanek, K.; Fenoglio, D.; et al. A new marine-derived sulfoglycolipid triggers dendritic cell activation and immune adjuvant response. *Sci. Rep.* **2017**, *7*, 6286. [CrossRef] [PubMed]

22. Morawski, M.; Schilling, S.; Kreuzberger, M.; Waniek, A.; Jäger, C.; Koch, B.; Cynis, H.; Kehlen, A.; Arendt, T.; Hartlage-Rübsamen, M.; et al. Glutaminyl cyclase in human cortex: Correlation with (pGlu)-amyloid-β load and cognitive decline in Alzheimer's disease. *J. Alzheimer's Dis.* **2014**, *39*, 385–400. [CrossRef] [PubMed]

23. Joyard, J.; Maréchal, E.; Miège, C.; Block, M.A.; Dorne, A.-J.; Douce, R. Structure, Distribution and Biosynthesis of Glycerolipids from Higher Plant Chloroplasts. In *Lipids in Photosynthesis: Structure, Function and Genetics*; Springer: Dordrecht, The Netherlands, 2006; pp. 21–52.

24. Benning, C.; Ohta, H. Three enzyme systems for galactoglycerolipid biosynthesis are coordinately regulated in plants. *J. Biol. Chem.* **2005**, *280*, 2397–2400. [CrossRef] [PubMed]

25. Kobayashi, K.; Nakamura, Y.; Ohta, H. Type A and type B monogalactosyldiacylglycerol synthases are spatially and functionally separated in the plastids of higher plants. *Plant. Physiol. Biochem.* **2009**, *47*, 518–525. [CrossRef]

26. Qi, Y.; Yamauchi, Y.; Ling, J.; Kawano, N.; Li, D.; Tanaka, K. Cloning of a putative monogalactosyldiacylglycerol synthase gene from rice (*Oryza sativa L.*) plants and its expression in response to submergence and other stresses. *Planta* **2004**, *219*, 450–458. [CrossRef]

27. Myers, A.M.; James, M.G.; Lin, Q.; Yi, G.; Stinard, P.S.; Hennen-Bierwagen, T.A.; Becraft, P.W. Maize opaque5 encodes monogalactosyldiacylglycerol synthase and specifically affects galactolipids necessary for amyloplast and chloroplast function. *Plant. Cell* **2011**, *23*, 2331–2347. [CrossRef]

28. Dubots, E.; Audry, M.; Yamaryo, Y.; Bastien, O.; Ohta, H.; Breton, C.; Maréchal, E.; Block, M.A. Activation of the chloroplast monogalactosyldiacylglycerol synthase MGD1 by phosphatidic acid and phosphatidylglycerol. *J. Biol. Chem.* **2010**, *285*, 6003–6011. [CrossRef]

29. Basnet, R.; Zhang, J.; Hussain, N.; Shu, Q. Characterization and Mutational Analysis of a Monogalactosyldiacylglycerol Synthase Gene OsMGD2 in Rice. *Front. Plant. Sci.* **2019**, *10*, 992. [CrossRef]

30. Awai, K.; Maréchal, E.; Block, M.A.; Brun, D.; Masuda, T.; Shimada, H.; Takamiya, K.I.; Ohta, H.; Joyard, J. Two types of MGDG synthase genes, found widely in both 16:3 and 18:3 plants, differentially mediate galactolipid syntheses in photosynthetic and nonphotosynthetic tissues in *Arabidopsis thaliana*. *Proc. Natl. Acad. Sci. USA* **2001**, *98*, 10960–10965. [CrossRef]

31. Frentzen, M. Phosphatidylglycerol and sulfoquinovosyldiacylglycerol: Anionic membrane lipids and phosphate regulation. *Curr. Opin. Plant. Biol.* **2004**, *7*, 270–276. [CrossRef]

32. Roy, A.B.; Hewlins, M.J.; Ellis, A.J.; Harwood, J.L.; White, G.F. Glycolytic breakdown of sulfoquinovose in bacteria: A missing link in the sulfur cycle. *Appl. Environ. Microbiol.* **2003**, *69*, 6434–6441. [CrossRef] [PubMed]

33. Speciale, G.; Jin, Y.; Davies, G.J.; Williams, S.J.; Goddard-Borger, E.D. YihQ is a sulfoquinovosidase that cleaves sulfoquinovosyl diacylglyceride sulfolipids. *Nat. Chem. Biol.* **2016**, *12*, 215–217. [CrossRef] [PubMed]

34. Sanda, S.; Leustek, T.; Theisen, M.J.; Garavito, R.M.; Benning, C. Recombinant *Arabidopsis* SQD1 converts UDP-glucose and sulfite to the sulfolipid head group precursor UDP-sulfoquinovose in vitro. *J. Biol. Chem.* **2001**, *276*, 3941–3946. [CrossRef] [PubMed]

35. Yu, B.; Xu, C.C.; Benning, C. *Arabidopsis* disrupted in SQD2 encoding sulfolipid synthase is impaired in phosphate-limited growth. *Proc. Natl. Acad. Sci. USA* **2002**, *99*, 5732–5737. [CrossRef] [PubMed]

36. Lauritano, C.; Martín, J.; De La Cruz, M.; Reyes, F.; Romano, G.; Ianora, A. First identification of marine diatoms with anti-tuberculosis activity. *Sci. Rep.* **2018**, *8*, 2284. [CrossRef]

37. Brillatz, T.; Lauritano, C.; Jacmin, M.; Khamma, S.; Marcourt, L.; Righi, D.; Romano, G.; Esposito, F.; Ianora, A.; Queiroz, E.F.; et al. Zebrafish-based identification of the antiseizure nucleoside inosine from the marine diatom *Skeletonema marinoi*. *PLoS ONE* **2018**, *13*, e0196195. [CrossRef]

38. Lauritano, C.; Andersen, J.H.; Hansen, E.; Albrigtsen, M.; Escalera, L.; Esposito, F.; Helland, K.; Hanssen, K.O.; Romano, G.; Ianora, A. Bioactivity Screening of Microalgae for Antioxidant, Anti-Inflammatory, Anticancer, Anti-Diabetes, and Antibacterial Activities. *Front. Mar. Sci.* **2016**, *3*. [CrossRef]

39. Riccio, G.; Lauritano, C. Microalgae with Immunomodulatory Activities. *Mar. Drugs* **2020**, *18*, 2. [CrossRef]

40. Giordano, D.; Costantini, M.; Coppola, D.; Lauritano, C.; Núñez Pons, L.; Ruocco, N.; di Prisco, G.; Ianora, A.; Verde, C. Biotechnological Applications of Bioactive Peptides From Marine Sources. In *Advances in Microbial Physiology*; Elsevier: Cambridge, UK, 2018; pp. 171–220. ISBN 9780128151907.

41. Keeling, P.J. The endosymbiotic origin, diversification and fate of plastids. *Philos. Trans. R. Soc. B Biol. Sci.* **2010**, *365*, 729–748. [CrossRef]

42. Falkowski, P.G.; Katz, M.E.; Knoll, A.H.; Quigg, A.; Raven, J.A.; Schofield, O.; Taylor, F.J.R. The evolution of modern eukaryotic phytoplankton. *Science* **2004**, *305*, 354–360. [CrossRef]

43. Manzo, E.; Gallo, C.; Sartorius, R.; Nuzzo, G.; Sardo, A.; De Berardinis, P.; Fontana, A.; Cutignano, A. Immunostimulatory Phosphatidylmonogalactosyldiacylglycerols (PGDG) from the marine diatom *Thalassiosira weissflogii*: Inspiration for a novel synthetic toll-like receptor 4 agonist. *Mar. Drugs* **2019**, *17*, 103. [CrossRef] [PubMed]

44. Zheng, Y.; Anderson, S.; Zhang, Y.; Garavito, R.M. The structure of sucrose synthase-1 from *Arabidopsis thaliana* and its functional implications. *J. Biol. Chem.* **2011**, *286*, 36108–36118. [CrossRef] [PubMed]

45. Rocha, J.; Sarkis, J.; Thomas, A.; Pitou, L.; Radzimanowski, J.; Audry, M.; Chazalet, V.; de Sanctis, D.; Palcic, M.M.; Block, M.A.; et al. Structural insights and membrane binding properties of MGD1, the major galactolipid synthase in plants. *Plant. J.* **2016**, *85*, 622–633. [CrossRef] [PubMed]

46. Mulichak, A.M.; Theisen, M.J.; Essigmann, B.; Benning, C.; Garavito, R.M. Crystal structure of SQD1, an enzyme involved in the biosynthesis of the plant sulfolipid headgroup donor UDP-sulfoquinovose. *Proc. Natl. Acad. Sci. USA* **1999**, *96*, 13097–13102. [CrossRef]

47. Okazaki, Y.; Otsuki, H.; Narisawa, T.; Kobayashi, M.; Sawai, S.; Kamide, Y.; Kusano, M.; Aoki, T.; Hirai, M.Y.; Saito, K. A new class of plant lipid is essential for protection against phosphorus depletion. *Nat. Commun.* **2013**, *4*, 1510. [CrossRef]

48. Zhan, X.; Shen, Q.; Wang, X.; Hong, Y. The Sulfoquinovosyltransferase-like Enzyme SQD2.2 is Involved in Flavonoid Glycosylation, Regulating Sugar Metabolism and Seed Setting in Rice. *Sci. Rep.* **2017**, *7*, 4685. [CrossRef]

49. Zhan, X.; Shen, Q.; Chen, J.; Yang, P.; Wang, X.; Hong, Y. Rice sulfoquinovosyltransferase SQD2.1 mediates flavonoid glycosylation and enhances tolerance to osmotic stress. *Plant. Cell Environ.* **2019**, *42*, 2215–2230. [CrossRef]

50. Altschul, S. Basic Local Alignment Search Tool. *J. Mol. Biol.* **1990**, *215*, 403–410. [CrossRef]

51. Villar, E.; Vannier, T.; Vernette, C.; Lescot, M.; Cuenca, M.; Alexandre, A.; Bachelerie, P.; Rosnet, T.; Pelletier, E.; Sunagawa, S.; et al. The Ocean Gene Atlas: Exploring the biogeography of plankton genes online. *Nucleic Acids Res.* **2018**, *46*, W289–W295. [CrossRef]

52. Keeling, P.J.; Burki, F.; Wilcox, H.M.; Allam, B.; Allen, E.E.; Amaral-Zettler, L.A.; Armbrust, E.V.; Archibald, J.M.; Bharti, A.K.; Bell, C.J.; et al. The Marine Microbial Eukaryote Transcriptome Sequencing Project (MMETSP): Illuminating the Functional Diversity of Eukaryotic Life in the Oceans through Transcriptome Sequencing. *PLoS Biol.* **2014**, *12*, e1001889. [CrossRef]

53. Carradec, Q.; Pelletier, E.; Da Silva, C.; Alberti, A.; Seeleuthner, Y.; Blanc-Mathieu, R.; Lima-Mendez, G.; Rocha, F.; Tirichine, L.; Labadie, K.; et al. A global ocean atlas of eukaryotic genes. *Nat. Commun.* **2018**, *9*, 373. [CrossRef] [PubMed]

54. Salazar, G.; Paoli, L.; Alberti, A.; Huerta-Cepas, J.; Ruscheweyh, H.J.; Cuenca, M.; Field, C.M.; Coelho, L.P.; Cruaud, C.; Engelen, S.; et al. Gene Expression Changes and Community Turnover Differentially Shape the Global Ocean Metatranscriptome. *Cell* **2019**, *179*, 1068–1083. [CrossRef] [PubMed]

55. Rusch, D.B.; Halpern, A.L.; Sutton, G.; Heidelberg, K.B.; Williamson, S.; Yooseph, S.; Wu, D.; Eisen, J.A.; Hoffman, J.M.; Remington, K.; et al. The Sorcerer II Global Ocean Sampling expedition: Northwest Atlantic through eastern tropical Pacific. *PLoS Biol.* **2007**, *5*, e77. [CrossRef] [PubMed]

56. Conesa, A.; Götz, S.; García-Gómez, J.M.; Terol, J.; Talón, M.; Robles, M. Blast2GO: A universal tool for annotation, visualization and analysis in functional genomics research. *Bioinformatics* **2005**, *21*, 3674–3676. [CrossRef] [PubMed]

57. R Core Team. 2019: A language and environment for statistical computing. R Found. Stat. Comput. Vienna, Austria. Available online: http//.r-project.org/ (accessed on 18 November 2019).

58. Wickham, H. Package 'scales'—Scale Functions for Visualization. *CRAN Repos.* **2018**.

59. Wickham, H. ggplot2 Elegant Graphics for Data Analysis. Springer: Dordrecht, The Netherlands, 2016; ISBN 978-3-319-24275-0.

60. Papadopoulos, J.S.; Agarwala, R. COBALT: Constraint-based alignment tool for multiple protein sequences. *Bioinformatics* **2007**, *23*, 1073–1079. [CrossRef]

61. Capella-Gutiérrez, S.; Silla-Martínez, J.M.; Gabaldón, T. trimAl: A tool for automated alignment trimming in large-scale phylogenetic analyses. *Bioinformatics* **2009**, *25*, 1972–1973. [CrossRef]

62. Guindon, S.; Gascuel, O. A Simple, Fast, and Accurate Algorithm to Estimate Large Phylogenies by Maximum Likelihood. *Syst. Biol.* **2003**, *52*, 596–704. [CrossRef]

63. Le, S.Q.; Gascuel, O. An improved general amino acid replacement matrix. *Mol. Biol. Evol.* **2008**, *25*, 1307–1320. [CrossRef]

64. Lefort, V.; Longueville, J.E.; Gascuel, O. SMS: Smart Model Selection in PhyML. *Mol. Biol. Evol.* **2017**, *34*, 2422–2424. [CrossRef]

65. Anisimova, M.; Gascuel, O. Approximate likelihood-ratio test for branches: A fast, accurate, and powerful alternative. *Syst. Biol.* **2006**, *55*, 539–552. [CrossRef] [PubMed]

66. Kelley, L.A.; Mezulis, S.; Yates, C.M.; Wass, M.N.; Sternberg, M.J.E. The Phyre2 web portal for protein modeling, prediction and analysis. *Nat. Protoc.* **2015**, *10*, 845–858. [CrossRef] [PubMed]

67. Yang, J.; Roy, A.; Zhang, Y. BioLiP: A semi-manually curated database for biologically relevant ligand-protein interactions. *Nucleic Acids Res.* **2013**, *41*, D1096–D1103. [CrossRef] [PubMed]

68. Yang, J.; Roy, A.; Zhang, Y. Protein-ligand binding site recognition using complementary binding-specific substructure comparison and sequence profile alignment. *Bioinformatics* **2013**, *29*, 2588–2595. [CrossRef] [PubMed]

69. McNicholas, S.; Potterton, E.; Wilson, K.S.; Noble, M.E.M. Presenting your structures: The CCP4mg molecular-graphics software. *Acta Crystallogr. Sect. D Biol. Crystallogr.* **2011**, *67*, 386–394. [CrossRef] [PubMed]

MDPI
St. Alban-Anlage 66
4052 Basel
Switzerland
Tel. +41 61 683 77 34
Fax +41 61 302 89 18
www.mdpi.com

Marine Drugs Editorial Office
E-mail: marinedrugs@mdpi.com
www.mdpi.com/journal/marinedrugs